DECISIONS IN DOUBT

THE NELSON A. ROCKEFELLER SERIES

in Social Science and Public Policy

Brian J. L. Berry, 1992
America's Utopian Experiments: Communal Havens
from Long-Wave Crises

Giulietto Chiesa, with Douglas Taylor Northrop, 1993
Transition to Democracy: Political Change
in the Soviet Union, 1987–1991

Wade L. Robison, 1994
Decisions in Doubt: The
Environment and Public Policy

WADE L. ROBISON

Decisions in Doubt

THE ENVIRONMENT
AND PUBLIC POLICY

DARTMOUTH COLLEGE
Published by University Press of New England / Hanover and London

Dartmouth College
Published by University Press of New England, Hanover, NH 03755
© 1994 by the Trustees of Dartmouth College
All rights reserved
Printed in the United States of America
5 4 3 2 1
CIP data appear at the end of the book

CONTENTS

About the Series   vii
Preface   ix

1. Making the Best Choices   1
   The Way We Make Decisions   1
   Our Ongoing Practices   5
   Who Makes Decisions?   6
   What We Face and Where We Are Going   8
   What to Keep in Mind   10
2. Natural Social Artifacts   14
   Natural Artifacts   15
   Artifacts and Intentions   19
   The Scope of Natural Artifacts   25
   Normative Artifacts   28
   Morality and Values   32
3. Knowledge and Doubt   35
   The Epistemological Card   38
   Deciding Reasonably without All Relevant Information   41
   The Situation We Are In   47
   The Best That We Can Do   52
4. Who Decides?   57
   Professional Positions   58
   Political Decision Making   62
   Who Decides What   82
5. The Options   87
   Why Incineration?   89
   Focusing on the Long Term   96
   Changing Habits and Practical Problems   98

6. What Ought We to Do? 101
   Assessing Natural Social Artifacts 103
   Taking a Risk 107
   Clear Cases 109
   The Decision-Procedure 114
   So What Is New? 118
   Presuming an Accident 124
   A Careful Cast of Mind: Easy Cases 129
   The Relevance of Rights 135
   Other Rights and Interests 143
   How to Proceed 145
7. A New Cast of Mind 148

   Notes 165
   Index 255

ABOUT THE SERIES

In the last decade of the twentieth century, as the global community faces a growing array of complex and interrelated social problems, public policymakers increasingly will turn to scholars and practitioners for guidance. To chronicle and disseminate the substance of the compelling discussions that will result, the Nelson A. Rockefeller Center for the Social Sciences at Dartmouth College, in collaboration with University Press of New England, has inaugurated this series of books.

Rockefeller Series books will be disparate in content but united in a common approach: presenting ways in which social scientific expertise is brought to bear on public policy issues of current or historic importance. The specific topics addressed will be as diverse as were the interests and work of U.S. Vice-President Nelson A. Rockefeller, which included state and local government, the environment, Third World economic development, publicly funded art, racism and intergroup conflict, and the functioning of communities. Authors will assess historical or existing policies, as well as the need for new or adjusted policies, in a search for viable solutions to pressing social, political, and economic problems.

The Rockefeller Series draws upon two sources for its books, the annual Nelson A. Rockefeller Monograph Competition, and works generated from Rockefeller Center research programs or Center-sponsored conferences. Included in the latter are scholarly works originating in one or more of the eight social science departments associated with the Center.

The overriding goal of the Nelson A. Rockefeller Series is to stimulate academics, policymakers, practitioners, and the public to think about and understand societal processes and the public policy implications associated with them. It is our fondest hope that these volumes will promulgate innovative and useful ideas, for as P. W. Bridgman notes, "There is no adequate defense, except stupidity, against the impact of a new idea."

George J. Demko, Director of the Rockefeller Center

PREFACE

In the mid-1970s, the State of Michigan found itself with an environmental crisis of the first magnitude. Polybrominated biphenyl (PBB) had accidentally been mixed into animal feed and entered the food chain. I talked and wrote about the problems and found myself generally frustrated by the discussion. Those of us concerned to enter into it had to engage in an analysis of the various scientific issues involved and make sure that we had all the facts straight. Yet, we then entered a land, I felt, where little agreement and almost no chance of consensus could emerge: The energy of an enormous number of engaged citizens, politicians, and scientists went in, and little came out.

The inaction seemed sanctioned when we tried to assess what harm, if any, was caused by the introduction of PBB into the food chain. We then had to project the consequences for human health of a substance, PBB, not yet subjected to experimental investigation. Nothing certain could emerge. So, it was argued, certainly nothing should be done. For, after all, although we did not know that PBB was harmful, we did know that acting to remove it completely from the environment would be harmful to a variety of interests, economic interests in the state not the least among those. The implicit appeal was to a prima facie plausible model of how rational persons ought to act: When we are faced with options, one of which clearly causes harm and another of which may not cause harm, we act rationally only to the extent that we act with knowledge and act to avoid certain, rather than merely possible, harm.

I was struck by the way we tended to think about that public policy issue. Thinking about how we think about something is a philosophical enterprise par excellence, and in thinking about our modes of thought on public policy matters generally, I found that we seem condemned, in order to be what we take as being rational, to privilege the status quo as we wait for the knowledge that we need to be sure we know exactly what we are doing. We privilege the status quo, even if it may cause great harm to do that, and we risk such harm even though we could act to diminish potential risks and do so in ways that are not especially costly.

I found that puzzling, and this book is an attempt to work out the puzzle and suggest that what we need is a different way of thinking about some public policy issues and environmental issues in particular. We need a paradigm about how to make reasonable decisions, one that not only itself makes sense but also does not condemn us to inaction when confronted with risks of great magnitude and when we doubt that we have all the knowledge we could have about an issue.

This is a book on how we ought to reason about public policy matters, and I want it to affect how we make public policy decisions. One worry readers may have is that they will have to enter too deeply into the thickets of philosophical disagreements, wondering where the thrust of the discussion has gone. But, although I think that theoretical issues of great complexity and importance always hover in the background of public policy discussions, I have tried to bring in only what is essential for the argument I am making. What I lose in examining thoroughly the theoretical underpinning, I hope to gain in clarity of presentation.

I have tried to make the sources of my work accessible to all. It is frustrating, even to those with access to major libraries, that many of the books and articles on which the discussion of major issues seems to rely are either difficult or impossible to obtain or, if obtained, virtually inaccessible to the general reader because they are so heavily laden with scientific terminology, some of which is necessary for research scholarship. I have thus made heavy use of newspapers, especially the *New York Times*, in part because such sources are readily available and in part because the discussions they contain are often significantly more accessible than the scientific reports from which they draw. So I have sometimes purposefully chosen not to quote an article from a scientific journal when a report of the article in the newspapers or elsewhere is clearer.

But a more important reason for citing daily newspapers and news magazines is that it is in their pages that we can find most clearly the conflict between various ways of making decisions about public policy matters, as they report the often unselfconscious reactions, pro and con, to various new discoveries and to the many conflicts that arise about public policy matters.

I owe much to many and have given little indication of my intellectual sources. I am not sure I could provide the sources of much of my thinking about the decision-procedure I think we should use, for instance, or for my sense of our epistemic state regarding such matters. For my thinking has been stimulated by reading all kinds of material and by all sorts of discussions with friends, colleagues, and students.

One example may suffice to explain my difficulties in acknowledging my debts. I became convinced of the power that a form of thought has over our capacity to understand and solve a problem by Bernard Bailyn's response to a puzzle about the American Revolution.[1] The various British acts that led to the Declaration of Independence may have been a series of independent miscalculations, but the Americans assumed a conspiracy on the part of the king's ministers. Why? Bailyn's explanation turns on the dominance of a mode of thought at the time: acts that could be the result of intention are the result of intention. The series of taxes and impositions and other acts could be tied together and so, by this mode of conspiratorial thinking, were tied together—if not by the king, then by Parliament, and if not by Parliament, then by others. Rousseau's *Confessions* is a wonderful, though depressing, example of this mode of thought. Adam Smith's analysis of how such ends as providing for the greatest economic happiness for the greatest number can be explained by the acts of individuals intending only their own benefit provides a completely different mode of thinking about such matters. Such are my sources. I hope I will be forgiven my failure to make them all clear.

I owe much to Richard Cook, a colleague at Kalamazoo College in the Chemistry Department who served on former Governor Blanchard's Michigan Toxic Substance Control Commission and wrote a monograph on the problems of incineration.[2] Reading his work convinced me to pursue the subjects the PBB crisis had brought to my attention, for I found him puzzled, in regard to incineration, about some of the same problems I puzzled over regarding the PBB crisis.

I should make clear my thanks to the French and Portuguese postal systems, both of which lost a first draft of this manuscript and saved me from the embarrassment of having that draft critiqued by others, and to Leslie Francis, who has read this manuscript in various stages of disarray with such patience and encouragement that I shall never be able to repay her. I also need to thank the reviewers of this manuscript, Hugo Bedeau, Kent E. Portney, and a third anonymous reviewer, all of whom helped me immensely in clarifying my argument. They undoubtedly still will disagree with some of my contentions, and with the inadequacies of my responses to their concerns, but I think their target is at least clearer, thanks to them.

Pat and Bill Hale provided the funding for the Ezra A. Hale Chair in Applied Ethics, which I have held these past few years, and I thank them for providing the resources for me to be able to do this work. I should thank as well the National Endowment for the Humanities, for an initial draft of this project was sketched out sitting at a café on the Cours Mirabeau in Aix-en-Provence where I had gone to write on David Hume while on an NEH Fellowship.

One does not write in an intellectual and emotional vacuum, and I thank Pamela Grath, Chris Latiolais, Kit Mayberry, Jane McIntyre, Michael Pritchard, Linda Reeser, David Scarrow, and numerous others for their friendship, their conversation, and their caring in times and situations not of their choosing. If friendship had a currency and allowed for debts, mine would be great indeed. To Kit I owe too much to measure.

I dedicate this to my children, Kelly and April, and to Kit's children, Carrie and Megan, whom I hope will come to live in a world in which this and similar works were not a waste.

W. L. R.
July 1994

DECISIONS IN DOUBT

CHAPTER ONE

Making the Best Choices

Anyone who wants to make the best of all possible choices regarding the environment is at a disadvantage. We start with an accumulation of waste even God might have difficulty transmuting into the goods of paradise, and waste is just the most visible of problems regarding our environment—from chlorofluorocarbons in the atmosphere causing a depletion of the ozone, to lead in our water supplies, to various chemicals and compounds in the food chain. Any decisions we make must respond to all we already have: It is too late for Eden.

The temptation is to chuck everything and start afresh, but, however understandable, that sort of impulse has helped lead us into this mess. Besides, as we come to investigate how we make decisions about the environment, who makes them, and what sort of situation they find themselves in as they make them, we shall see how difficult it is to chuck anything at all. Part of our problem is that we carry an accumulation of intellectual and institutional baggage that impedes the very way we think about how to respond to environmental problems. If we are not to make matters worse—let alone have a chance to regain even a portion of Eden—we need to change the way we think about environmental matters. We can best see how by looking at an example of the way in which decision making regarding environmental matters tends to proceed.

The Way We Make Decisions

In the mid-1970s the Michigan Department of Agriculture and Department of Public Health were faced with the issue of whether to ban all foodstuffs tainted with polybrominated biphenyl (PBB)—from beef to pork to chickens to eggs. Polybrominated biphenyl is a relative of PCB, the chemical dumped in Japan that caused such damage to fetuses, and it was discovered in mid-1974 that PBB had accidentally been mixed with animal feed in 1973. The Department of Agriculture moved relatively quickly to ban the sale of milk adulterated with a level of PBB it considered dangerous, but the official response slowed quickly.

No research had ever been done on PBB's effects on humans, and no evidence was available as to whether PBB is harmful to humans. It killed cattle, even when ingested in small amounts, but we did not know its impact on human health. In addition, the damage to the state's economy and to particular persons and companies in agriculture would have been serious if all tainted foodstuffs had been banned. Disagreements arose about what amount of PBB was enough to provide a basis for action.[1] But what slowed the movement to ban the sale of all foodstuffs with PBB was a powerful argument about what is necessary to justify public policy decisions of such magnitude.[2]

Because we lacked knowledge of PBB's actual effects on humans, it was argued, it would not be rational to ban those foodstuffs. We would have to ban them without knowing they were harmful, and yet banning them would itself cause harm.[3] It would not be rational, it was presumed, to cause such harm without knowledge of harm.[4]

The PBB in the environment would long since have been diluted beyond the reach of measuring instruments before any medical studies of its short-term or long-term effects on human health could be conceived, funded, and brought to conclusion. So the demand for a showing of even some small harm to human health effectively blocked action to mitigate whatever harm might have been occurring.

Behind that demand, I argue, is a mistaken presumption of what is required for a fully rational decision, a presumption that all too often powers public policy decisions regarding the environment. It does not help that the presumption is one we all find so plausible that to call it a presumption rather than a requirement for rational decisions will, it may be thought, already beg some question or other. For the presumption is that a fully rational decision requires all relevant information. Without all relevant information, it is presumed, we come to a rational decision by luck, and the less information we have that is relevant to the decision, the more luck is required to get it right, and so the less rational it is likely to be.

More than full information is required by this presumption, of course. The ideal is a godlike objectivity as well as omniscience. We humans will always be fallible, moved by passion, and deflected from correct decisions by bias, and, so, it is presumed, we must strive to overcome these handicaps by obtaining as much purported knowledge as we can about the issues and winnowing claims to knowledge through the scientific method.[5] Those who make rational decisions ought not to be moved by bias or passion but by what reason requires; by this presumption of what is rational, reason requires all relevant information for a fully rational decision. Without all relevant information, even a god, who is presumptively objective, can only make the correct decision by luck.

We, who can make no pretense to such objectivity, ought to be even more careful, it follows, to be sure not to base any public policy decision on less than all relevant information. The recent finding of a correlation between soot and upward to 60,000 deaths a year might be thought to justify action to mitigate the potential harm caused by such soot, but the correlation may be the result of other undiscovered factors, and so, it is claimed, "We shouldn't make major policy decisions about causes of death based on the kind of evidence we have so far."[6] In short, this presumptive vision of how to make public policy decisions tells us that we need further information if we are to have enough to justify any action.

This presumptive vision has its place, as I argue, but in fact we are never in a position to know, regarding any environmental matter of any moment, that we have all relevant information. Requiring that we have all relevant information thus sets an heuristic ideal, powering continual demands for more information and thus for more research. Because the ideal is that we have *all* relevant information before acting, it stymies any action before that information becomes available. "All relevant information" is a moving target—frustrating for someone concerned that such public entities as the Michigan Department of Public Health act to prevent harm that may be occurring, but capable of being used to frustrate anything that might change the status quo, from a waste incinerator in the neighborhood to a factory down the road.[7]

Yet, the plausibility and power of the ideal are difficult to deny. Questions about the impact on human health of chemical compounds such as PBB or those in toxic wastes are preeminently scientific, and because there are facts to be known about what harm, if any, those compounds cause, how could any rational person, one wants to ask, make rational decisions about such compounds without knowledge of such facts? What I am calling a presumption looks to be a requirement for rational decision making.

It thus powers a set of procedures as well as furthers, in some cases, the political paralysis that tends to occur when momentous public policy matters that affect different constituencies in different ways are at issue. When faced with a problem, we are required, because of this presumption, to gather as much information as possible before we can act—Is the ozone layer really being depleted, and if so, to what extent, how quickly, and why? Is there acid rain? If so, is it causing environmental damage, how is it produced, how extensive is the damage, and how quickly is it occurring? The presumption is that we must wait until the evidence is in before making a decision.[8]

Banning asbestos in the 1930s, for instance, when its commercial use for insulating and roofing exploded, would not have been rational given this presumption because, although we had some evidence even then of its hazardous nature, we could not have known the full extent of its hazardous effects. Asbestiosis can be a long time in coming, manifesting itself only after twenty to thirty years, and the extent of damage to the lungs varies "directly with the duration of exposure."[9] So banning asbestos in the 1930s would have guaranteed that we would have incomplete information about what percentage of those working with it would be harmed and incomplete information about how much harm would occur to those harmed. Of course, waiting for more than thirty years to assess the exact harm precluded preventing it.[10] Waiting for all relevant information meant that we let the current practices continue, to the detriment of those exposed to asbestos and at great subsequent cost for its removal from our environment. So the natural presumption about how to make a rational decision is not free of costs.[11]

That this natural presumption has those sorts of costs is a reason to ask whether it is being made regarding that most visible of our environmental problems, waste, and whether, given those costs, it is the most appropriate presumption to make. We have another reason to ask as well. Our current practice is being determined without all relevant information. Waste is being pumped into deep wells, landfills are being filled, incinerators are being built, and trash and garbage burned—all without all relevant information. Waiting until we have full information to make what are presumptively rational decisions means that we shall continue to make decisions, while we wait, without all relevant information and that we shall continue to live with the accumulated fallout of past decisions made without all relevant information. The capital investments we continue to make in purification plants, transportation systems, incinerators, and landfills could very well make acting on all relevant information difficult, if not politically and economically impossible, if we ever achieve it.

That is not a rational state of affairs, and we need to ask if it is rational to presume a vision of how we ought to make a rational decision that prevents us by its standards from now making rational decisions—especially when the consequence is that, while waiting for all relevant information, we continue making decisions that may cause great harm. Presuming this vision of how to make choices guarantees that we will not make the best of all possible choices.

Our Ongoing Practices

We have another difficulty in making the best of all possible choices regarding the environment. Consider waste again. We have in place a waste management policy achieved through years of decisions, most of which seemed rational at the time they were made. We must locate any new practice in the midst of this ongoing practice, and the current practice can be changed not by a new practice, equally good, but only by one better enough to justify the disruption to current practices change would bring.

The problem we face is similar to that created by a precedent in the law. A judicial decision sets the law, and citizens will rightly shape their affairs on the basis of that decision. If a court decides a case involving a municipality's zoning regulations in a way that insulates some residential property from surrounding commercial development, then it is reasonable for citizens to opt to invest in ways tailored to the court's decision. If the investment is for the long term, as purchasing a house tends to be, citizens have at least *some* right to expect the law to be for the long term as well. If we invest our money in a house, deciding that it will at least retain its value for resale because nearby commercial development is not permissible, we will be harmed substantially if the law is changed, and we will rightly feel more than chagrined if the change cannot be justified by a compelling reason.

It may seem a corollary of the truth, "If it ain't broke, don't fix it," that if it is wrong, we should change it. But to overturn a previous decision merely because it is wrong is to ignore the disruption to citizens' affairs that would occur. We need to add the accumulated weight of past practice, based on the prior decision, to the calculation about whether to overturn a previous decision. The new decision cannot then just be marginally better than the old. We must have a significantly better decision to justify the disruption to expectations the prior decision created.

Similarly, we have a practice of waste management: Citizens have habits of use, companies have invested money for producing and packaging goods and have practices of disposal, various governmental bodies have ordinances and laws and investments of money in plants and equipment. The solution we might choose if we were to decide to begin afresh, independently of our current practice, may so disrupt those practices and the expectations they engender as to be unjustifiable. So our current practice of waste management may get in the way of the best of all possible choices regarding waste.

At the minimum, our current practice may act as a drag on our decision making, deflecting us from the best choice by an understandable concern to wend our way carefully past already existing interests or to minimize disruptions.[12] At the maximum, our current practices may prevent us from taking seriously, as reasonable alternatives, suggestions for change, for, as I have suggested, our current practices include a presumption about how to make decisions that, I shall argue, excludes the usual form of argument marshalled by those concerned about our current practices and our environment.

Who Makes Decisions?

Yet, even if we work out how the most rational decisions about environmental matters ought to be made, given our current practices, such decisions will tend not to be made by those who, in our system, must make public policy decisions. If we are not to have complete laissez-faire decision making, in which those with any interest in waste management—consumers, waste hauling companies, landfill operators, manufacturers, bottlers, and so on—make whatever decisions they wish, then we will have decisions made by those charged in our system with setting public policy—not God, but elected officials, those appointed by elected officials, and those in civil service positions in federal, state, and local governments. Such decisions will be made by governors, legislators, county commissioners, city councilors, administrators in departments of public health, and so on.

A thorough sorting of these various classes would mark out differences, great and small. Someone on a county commission, who must face the electorate in hearing after hearing, may well react differently than a more insulated elected official. An administrator may respond in a different way than an elected official to a politically sensitive problem, worrying less about the reactions of the electorate than about maintaining the bureaucracy's reputation for neutrality, for instance.[13] Important as these differences are, however, we may ignore them for our purposes here, for sooner or later decisions about waste management tend to pass through the hands of elected officials, such as those on a county commission, who generally face reelection, if not this year, then next, and if not next, then soon after. We can thus distinguish and grade the sorts of decisions politicians face on the basis of their power to help or harm electoral chances.

The best sort of decision for any elected official to make is one in which all individuals and groups affected by the decision find their interests fully met. Such situations are rare and hardly ever reach the state where elected officials need to act on them, but when they do, such an official is in wonderland. For it is wonderful to be able to tell all one's constituents that all their interests have been fulfilled.

The next best sort of decision is one in which the elected official can argue, cogently and reasonably, that although not everyone's interests were satisfied, the decision was for the public good. The best variant of such a decision is one in which each party to the decision is compelled to admit that, although their own interests were not met or not fully met, the decision was truly best for all concerned. Politicians always want to argue that their decisions are for the public good, and decisions can be distinguished and graded within this arena by how truly they are, or are perceived to be, for the public good.[14]

We could go on to distinguish within this category further categories, for the ways in which acts can affect the public interest are many—depending on whether rights or merely interests are affected, whether groups or individual actions are most affected, and so on. But we do not need an anatomy of the variants of this species to recognize an elected official's interest in maintaining that a particular act is the best that could be achieved.

The last best alternative for elected officials is that the decision not be made by them at all, but by experts, usually scientific, who must make the decision because, it can reasonably be claimed, the matter is an issue for experts. In this sort of case the politician can claim political immunity from the decision and its consequences and turn upon anyone who is dissatisfied the query, "Would you prefer that the decision be made by someone like me who doesn't know what they are doing?"

In the first sort of case, no one's ox is gored. In the second, everyone's may be gored, but in a manner everyone is willing to tolerate and, in the best of cases for the politician, even applaud. In the last the elected official does not do any goring at all. Indeed, if elected officials make the decision, or are ultimately responsible for it, they may well find their oxen gored, no matter how appropriate the decision, and so they will be understandably reluctant to make any decision at all.

More bluntly, the structure of politics will make some answers regarding the environment preferable to others for a person acting in a political capacity. To say this is not to impugn politicians—to question their motives or intentions or capacity for good acts—but to claim that even the most well-intended person, acting from the best of motives, with a clear-headed understanding of what is morally acceptable and what is not, will, when acting in a political capacity, tend *not* to make the best of all possible choices, but, at best, the best *political* choice. The process has a design that *provokes* such choices—even if inappropriate to the sort of problem being resolved, even if clearly mistaken, short-sighted, and so on.

The continuing vision of upcoming elections produces a tendency not to make any decision at all but to postpone a decision until after the next election, after "a proper study" can be done. It also produces a tendency to make a decision, for example, in which the bad effects are far removed and the good effects more immediate, that is, before the next election rather than after. Again, the desire not to be held accountable for a decision's bad effects affects the very way the problem is perceived, as scientific, and produces a tendency to give the problem to experts, a tendency encouraged by the political benefit of having someone else decide.

What We Face and Where We Are Going

Leibniz has an image that may help make my point clear. He pictures God with all possible worlds in mind, deciding which, if any, to create. Because God is good, He will choose the best of all possible worlds, Leibniz argues, and because He is also all-powerful and all-knowing, He can create that world. Neither ignorance nor impotence skew His choice.

We, on the other hand, because we are fallible, cannot even be sure we have pictured all the possible choices regarding any environmental problem, but even if we got them all, our very way of thinking about the problem would skew our choice. Requiring all relevant knowledge before we can act, and concerned not to harm existing interests without a showing of greater real harm, we immunize those interests, even if they may be harmful, and risk more harm as we await more knowledge. Yet, even if we thought about environmental issues properly, the decision process we must use to make environmental decisions is not the process God invokes, on Leibniz's view, of simply choosing the best. We must rely on our political process, and although it is a process well-designed to handle certain sorts of problems, it is ill-designed to respond, among other things, to problems that, like many environmental issues, require heavy upfront burdens with the benefits far removed in time or, at least, beyond the next election. Rather, it is what we may call an error-provocative design.[15]

We thus face at least three problems besides our fallibility in making the best of all possible choices about the environment: a model of decision making that demands all relevant information and, so, tends to delay decision making, even at great cost now as well as later; an ongoing practice that may constrain our choices; and decision makers constrained by features of their own position that are, at a minimum, not designed to produce the best decision.

Each of these factors—who makes a decision, under what conditions it is made, and how it is to be made—needs to be considered at greater length. For all three affect the decision-making process and may skew the result as we try to make the best of all possible choices. We also need to consider that choice—what is rational and moral. For it will do us little good to examine how we may fail to achieve our end, or how we may need to compromise to come as close to that end as we can, if we have no clear idea what that end may be.

In the next chapter I consider the nature of our current system of using the resources of the earth. Our system is what I call a natural social artifact. Another example is a natural language, such as English. Natural social artifacts come about over time and so are generally subject only to change of certain sorts. New words are added to English constantly, for example, but the grammatical structure resists change. Similarly, although we readily add to the ways we pollute our environment, our system of using the earth's natural resources will tend to resist any but incremental change.

One feature of our current system is a deep-seated conception of how to make rational decisions about environmental matters and waste in particular, the subject of my concern in Chapters 3 and 4. I consider, respectively, the model of decision making we are inclined to use and the implication of the decision's being made by those in our society charged with making public policy decisions. These two chapters may make for depressing reading, for they concentrate on difficulties built into the decision-making process. They are meant, however, to provide a realistic conception of how we are situated by focusing on systemic features of our practice.

We will quickly find ourselves in the midst of a great deal of detail in these chapters, and it is worth noting, before we immerse ourselves, that when faced with any problem, we are most likely to ignore those features which form our practice, for it is those to which we are most accustomed and which seem most natural to us: We tend not to notice them, and if we do, we tend to think of them as permanent features of the situation, immune to change. The more systemic those features are, the more permanent and so less subject to change they seem to be. They blend into the background against which we see the problem. Pulling them out to note their full effects may be depressing, for many cannot be changed without radical alterations in our whole way of moving through the world.

But by noting these systemic features of our practices we can at least guard ourselves against them. The end to be achieved in these chapters is not pessimism but a properly tempered and realistic optimism. Only if we understand the lay of the land will be we able to move through it expeditiously and have any hope of achieving any version of Eden.

In Chapter 5 I turn to the way we conceive of the problem regarding waste, as one we face rather than as one we create by our way of living in this world. The point is to make clearer our options. In Chapter 6 I examine the ends we ought to achieve, considering, at the same time, how best to get from where we are—with all the waste we have and the policies and practices and habits we now have in place—to where we ought to be, with the least amount of waste.

The focus of my concern is on how we conceive of making decisions regarding the environmental problems we face. In Chapter 7, I summarize the change I am arguing for in the way we ought to think about making those decisions. By the time we have reached that point, what I say will seem so obvious, I hope, as to be hardly worth saying.

I use as my running example the problem of waste. A problem endemic to examples is that we must be cautious about generalizing, but our problems with waste are, I think, paradigmatic for our environmental problems. What holds for how we ought to make decisions regarding it thus provides us with a model for how we ought to make decisions regarding other environmental issues. I have provided a variety of cases drawn from other environmental concerns to make that point. But I have also chosen to concentrate on waste because it is the environmental problem that has the most visible impact on our daily lives. It is thus the environmental problem most likely to engage our attention.

What to Keep in Mind

We should not think, however, that anyone can pull a wonderful and compelling solution to our environmental problems out of a magical hat. We must keep in mind that we need to change not only our way of managing waste but also the very way we *think* about our waste management practices. To the extent that we commit ourselves to the ideal vision that fully rational decisions require all relevant information, our current thinking and practices are incoherent—as we proceed to make often expensive decisions about waste *without* full information. We cannot be rational in such a situation, and so we cannot expect any solutions, let alone an easy and compelling one, without altering the way we conceive of our problems. This is not easy.

We cannot expect any more precision than the subject matter admits either,[16] and I argue that it admits at best some general principles about how to proceed, some suggestions about how we may easily be led astray, and a range of strategies that may be effective and will not bury us more deeply in waste.

The guiding principle that animates this discussion is that we can live in a more reasonable and better world than one in which garbage barges can bump from country to country for months looking for a dumping ground and in which waste haulers in the northeastern United States can seriously consider shipping waste to Cornwall, England.[17] Such scenarios are powered by laissez-faire economic considerations that fill the vacuum created when more reasonable solutions are wanting, and they are the stuff of jokes—late night television one-liners. We need not live in such a world.

The world we live in has a variety of kinds of waste—from radioactive material to industrial waste to air pollution to the ordinary contents of the waste basket in one's kitchen. There is no doubt that much of this is hazardous, likely, that is, to cause us physical harm.[18] Yet, it is a mark of how contentious the subject is and of how so little certainty there is that it is a constitutional issue to be decided by the Supreme Court, "whether the ash left from the burning of municipal solid waste must be treated as hazardous waste."[19] There seems no doubt that municipal solid waste contains what we would normally consider hazardous substances, such as lead from batteries chucked in the trash, and that incinerating such waste will leave much of this hazardous material as ash. But incinerators that produce energy are, by law, not to be treated as hazardous waste facilities. It was left unclear by the law whether the ash of such incinerators is to be treated as hazardous waste, but if it is, the incinerators would no longer be cost-effective. The issue is of no small importance because there are "137 municipal energy-recovery incinerators in the country, with dozens more being planned," and Chicago, where this case originated, now pays Michigan $23 a ton to bury the ash in an ordinary landfill and would have to pay "about $210 to store the ash as hazardous waste."[20]

So, if we tried to keep in mind hazardous waste as the object of our concern, we would begin with an idea itself confused by issues that plague decisions about waste management. Why, one might well ask, ought a determination of what is hazardous to our health be affected at all by a consideration of the cost of its disposal? It would seem to make more sense to determine what is hazardous based on what causes harm to human health and then decide whether to risk the harm by ordinary landfilling or mitigate the potential harm by special handling. But to ask such a question and to make such a judgment are already to begin to argue about how we ought to change the way we think about waste management. I prefer us to start, if we can, on more neutral territory.

But the terrain is not any less difficult. What we read carries different implications depending on what examples tend to pop to mind, and what I have to say will be more readily accessible if we tend to keep in mind the ordinary trash and garbage of our households—more the solid waste of municipal refuse[21] than the radioactive waste of reactors or the hazardous waste of industries, mines, and so on.

Much hazardous waste is the output of factories in which it is generally possible to determine, ahead of time, what compounds are in the waste and also possible to control both the production of such material and its disposal.[22] It is true that such waste is for the most part not now identified and controlled. The cruise of the *Karin B*, carrying hazardous waste originally from Italy and dumped illegally in Nigeria, is a metaphor for the lack of political will. Its voyage from Nigeria to England to the Netherlands to France and finally back to Italy where the seaport at which it was to dock passed an ordinance prohibiting its entry is a metaphor for the convoluted meanderings of the political mind.[23]

Yet, unlike our ordinary garbage and trash, hazardous waste from factories, mines, and so on presents a comparatively clean, if I may use that word, target for action: We can more readily identify it and isolate it, at least. More than that, the relevant political bodies are readily identifiable—national legislatures that regulate companies within their borders, with the assistance of such international bodies as the European Economic Community. So, although hazardous waste from such sources presents a problem, it presents a tidier problem than solid waste.

The same is true of radioactive waste from nuclear plants. It too presents a relatively crisp target for action—although as anyone familiar with the problems of what to do with such matter knows, the clarity of the target has yet to produce much clarity about what to do. Yet, if we can handle the solid waste of municipalities, we can surely begin to get a handle on waste that presents a tidier problem.

That is why I prefer the reader keep in mind the everyday garbage and trash of modern life—the sort of waste of concern to a county commission, a city council, or whatever other local political institution is responsible for decisions about waste management. I mean to make it clear just how intractable a problem waste may seem to those who, in our society, end up having the problem dumped on them.

Decisions about how to dispose of waste, made with the best of intentions, may come back to haunt those who must live with the harmful residue of waste that was thought harmless at the time.[24] Those now facing such decisions must keep in mind that others before them made decisions they thought good and reasonable that are now causing harm and costing enormous expense.[25]

Perhaps even worse, especially in those cases where it is clear that the waste is laced with hazardous and radioactive material—such as lead from old paint or batteries, or radium from old watch dials—it is unclear what ought to be done. We must intervene somewhere in the chain of the existence from useful product to waste, but it is unclear where (at the point of its production, for instance, requiring it to be made differently), unclear how (by controlling its disposal by requiring a refundable deposit, regulating its production so as to remove the hazardous material, and so on), and unclear who can and should intervene (local authorities who may have difficulty regulating products with more global markets, federal agencies that are sometimes difficult to move, and so on). What is usually clear is that the problem presents itself *as* a problem only long after decisions have been made that have made it a problem or aggravated its features. By the time the trash is a problem to a political body charged with handling it somehow, the time for intervention in its production is long past.[26]

These issues can thus seem so overwhelming that in facing them those who must make decisions about waste are unlikely to think about how they think about making the decisions.[27] But our vision of how to make such decisions is an incoherent guide to action. For we must continue to do something about the waste we generate as this guide tells us that we cannot make rational decisions without all relevant information. In asking the reader to keep in mind his or her own trash and garbage as we proceed to work our way out of the incoherent way we now think about environmental issues and waste in particular, I make the less than subtle suggestion that the problem is one each of us ought to face.

CHAPTER TWO

Natural Social Artifacts

It is sometimes suggested that what stymies change in the way we use the earth's resources are entrenched interests—of lobbying groups concerned not to alter the way we produce energy, of corporations with deep capital investments in incineration procedures, or of governmental bodies and agencies so used to dealing with such entrenched interests and corporations and the way they see the problems as to be unresponsive to local interests and changing concerns.

It is thus no surprise, riding through the countryside, to find signs of protest against some big company about to impose an incineration plant on the inhabitants of what had been a peaceful valley. Such signs attack the government as well for allowing it. In Allenwood, Pennsylvania, such signs dot the sides of the road: "Farmland, not Wasteland, Ban the Burner"; "Burn the Governor and the Incinerator"; "Take a Breath, It may be your Last"; "Legacy: Loot and Scoot. 57 Violations in One Day in Utah! Do You Want Those People In Our Valley?"[1]

Or it is sometimes suggested that the way we use the world could change if only individual citizens would mind and mend their ways—as though what is needed is some sort of environmental purification rite for us all, a baptism in environmental concerns. There is a hint of that suggestion in my asking the reader to keep in mind ordinary day-to-day kitchen trash and thus to face daily the problems the accumulation presents.

I do not dismiss these sorts of suggestions, but they do not come close to coming to grips with the problems we face in altering our ongoing practices.[2] It is a mistake to suppose, for instance, that if only people or corporations would see the light of reason, they would change their behavior. That presupposes that reason speaks with a single voice and that the practices some want changed are somehow clearly unreasonable. To understand fully in what ways they are not, and how difficult it therefore is to alter them, we must understand how deeply embedded they are in what I call a natural social artifact—*natural* because it is responsive to natural needs and wants; *social* because what is created as a response to those needs and wants is a normative practice that structures the way a person within that practice even understands the issues; and *artifact* because it is a human creation, a cooperative enterprise that is the fruit of innumerable acts and omissions, individual and otherwise.

Any such artifact, I argue, demands of those within it that they satisfy its norms, and, as we will see in Chapter 3, one of the features of our ongoing natural social artifact regarding the way we use the resources of the world is a certain way of reasoning about environmental problems. For these reasons it is next to useless to exclaim that if people were reasonable, they would change their practices and no longer lobby, for instance, to protect those practices. Those who argue within those practices *are* reasonable—according to the practice in which they take a part and according to those to whom they are addressing their arguments. They are, that is, reasonable *to us*.

It is a mistake, however, to suppose that we therefore need some environmental gurus who can lift themselves above our practices and, with greater wisdom than we all have, tell us how we ought to live our lives.[3] The patronizing attitude is not likely to win as many converts as would be necessary to alter our natural social artifact fundamentally, and it is unnecessary. We often reason as we ought to regarding many issues on the environment.[4] The task is not to try to take the high ground of rationality but to explain clearly the conditions under which such reasoning is appropriate and to show that it is appropriate regarding many of the thorniest environmental issues, including waste management.

To understand fully how this is so, and the implications for any attempts to change our practice, we need to examine the idea of a natural social artifact in more detail.

Natural Artifacts

Our practice of using the earth's resources is an artifact because it is the result of innumerable individual, corporate, and public policy decisions. It is something we have created in the same way we have created our language, our system of commerce, or our particular rules of courtesy. It is thus to be distinguished from a natural object such as a rock, for instance, which is not our creation, not the result of any human actions or omissions.[5]

But natural social artifacts are not just any kind of artifact. They are marked in at least two ways that distinguish them from the sorts of artifacts, a hammer, a telephone, that readily may come to mind. On the one hand, they bear no clear relation to any specific intention or set of intentions, and on the other there is a sense in which they must come into existence as a natural evolutionary response to certain natural needs. Let us briefly consider the latter first.

Our various natural languages are paradigmatic examples of natural social artifacts, and the differences between them are clear examples of how the natural need to communicate produces *some* social artifact—for language is surely our creation—but not necessarily the same artifact. Similarly, our natural needs for health produce some health care system, even though the variations we find in systems throughout the world indicate how differently we human beings can respond to the same sorts of natural needs. That there are deep similarities in various health care systems, just as there are deep similarities in various natural languages, is no surprise. These artifacts come into existence in response to individual needs and are shaped by countless acts and omissions that are themselves responsive to a variety of both causal factors and intended ends. What does not work or is inefficient, for instance, may be less likely to be an object of choice in creating some natural social artifact than what does work and is efficient.[6] It is no surprise that various natural languages allow us to communicate.

We produce waste, and the need to rid ourselves of our waste is as natural as any other natural need. We need only consider the products of our defecation to see how various systems of waste disposal evolve naturally out of such needs, and if we consider other by-products of our lives, such as the dregs of our meals or the leftovers of our productive lives, we can understand how just the natural need for eating and working space requires some means of waste disposal.

Natural social artifacts are *natural* because, although they are clearly creations and so artifacts, their creation is predictable, given natural human needs. A group of human beings will come to communicate with one another somehow, and the mode of communication will be a language of some sort. A group of human beings will come to dispose of their waste products somehow, and the mode of disposal will be a method of waste management of some sort.

Yet, what we create when we create a natural social artifact is an entity that is unlike many other human artifacts. The objects we humans make are numerous and immensely varied. Some are the creation of a specific intention or set of intentions. I choose to take a photograph and then to develop the film and make a print so that the result captures just what it was I was looking for when I took the photograph. From beginning to end, the process is marked by intentional action,[7] and although mistakes may occur, and I may come to see more or less in the photograph than I had thought was there and, accordingly, modify my original intentions to accommodate the change, the result is still clearly *my* photograph, an object I have created, an object of which I can be proud or that can dishearten me.

But such natural social artifacts as language do not bear that relatively simple set of relationships to intentional action. Each act or omission, each choice of word or phrase, each omission or addition of a phoneme, may perhaps occur because each person intended some end, specifically related to the use of the particular word or phoneme, but not necessarily and not likely. In any event, even if persons in such a case acted or omitted to act with some end in mind, they certainly do not necessarily intend the same end or even any end at all that may be the cause of the practice coming into question. No one may have intended that "Leicester" become pronounced "Lester," but that is the effect of innumerable elisions.

We do not all intend to pollute the air when we drive our cars, for instance, or when we smoke cigarettes, and what we do generally intend in driving our cars or smoking no doubt has little if anything at all to do with using the earth's resources or disposing of the waste we generate. It would be an oddity to find someone smoking just to pollute, although countless acts of smoking or not smoking produce a practice that affects the quality of air. Indeed, our acts and omissions may not even have a coherent set of ends, even if they have ends. We may pronounce a word in a certain way—"pin" for "pen," for instance—that may run afoul of our attempt to communicate in some situations. In short, what generates a natural social artifact is not to be found by examining a set of intentions but a set of acts and omissions, whatever their intentions, if any at all, with their consequences, intended or not.

The world is filled with such artifacts. Consider, as a simple example, a family's practices regarding what to do with dirty dishes. I use the same glass and same cup all day long, rinsing them out when I am finished and putting them in a place readily accessible for the next time I need them. My parents use new glasses each time they drink, rinsing the used glasses and putting them to the right of the sink, ready to be washed and rinsed by hand in scalding water in the evening. A friend and her children put their glasses in the sink, unrinsed, to be put by her in the dishwasher sometime. Each of these brief descriptions begins to sketch out a practice, each created by the acts and omissions of the various participants.

In my case the practice was the result of a conscious decision determined as much by a concern for economy of effort and by an aesthetic preference for a particular cup as by a concern about the economical use of resources. But although in this case the relation between intention and practice seems fairly direct, not all natural social artifacts bear such clear marks of intentional action.

Whatever the exact interplay of intentional action and natural social artifacts, in short, they are best conceived as the result of an accumulation of acts and omissions.[8] The result is an artifact, an artificial entity that would not have been produced but for those various acts and omissions, an entity or practice that may bear little resemblance to anything anyone would design to respond to the need or set of needs the practice in fact responds to. For instance, my parents' practice may have been determined by health concerns: One risks germs by drinking from a glass that has been sitting with moisture in it. But I am unsure of the causes of their practice, and I suspect their practice is more like that of my friend and her children, which may well have begun with the children not knowing what to do with their glasses and with my friend just picking up after them, either without thinking much about it or after deciding that the effort of getting after the children was more trouble than the effort of cleaning up after them. In any event, the practices these brief descriptions suggest are different, and yet all are informed by the acts and omissions of the participants, which themselves may or may not be informed by self-conscious thought and may not be the practice each would *design* if asked anew how they would now handle their dirty dishes.

The power of such an intimate example is that it displays both the range and the power of the concept of a natural social artifact: We produce many such artifacts, all the time; they may constantly change under the pressures of new factors, a dishwasher, for instance;[9] and they permeate the details of our lives and structure the very forms of our thought and our reasoning without our having intentionally created them.

They may range over such minute details as the way we handle dirty dishes, with a wide variety of variants from family to family and from subculture to subculture, and they can be so pervasive and deeply seated within our social structure that it may take some effort of thought to back off and realize that one is under the power of an artifact that could be different. It is in his inability to make that leap, above his thoughts, as it were, that marks the joke about the Englishman who, upon returning home after a tour of the world, pronounced himself happy to be back where people talked the way they thought. Anyone who remarks on a family's particular mode of handling dishes risks a puzzled, and perhaps offended, "How else could we do it?"

I am making three different points here about natural social artifacts, and they need to be carefully marked. One is that they have such a power over us, are often so deeply embedded that even the clash of one with another—as is illustrated by the joke about the Englishman—may not dislodge them. The power is that of a norm that is constitutive of, in that case, the very way one thinks about the world.

A second feature is their scope. They can range over such mundane features of our lives as how we handle dirty dishes to such pervasive features as our language or our conceptions of justice.[10] The broader their scope through society at large, the less likely we are to perceive them as artifacts, the more permanent and, so, less susceptible to change they will be perceived to be.

The third is that they are not marked by any simple relation to human intention or design. For instance, they may not be responsive to the ends for which we might think them designed. We may intend to get rid of harmful material from our kitchen by putting it in a bag and putting the bag out for trash collectors only to discover that its toxic residue—the lead in the batteries we used in the kitchen flashlight, for example—makes it way back in the water from our tap, polluted by leakage of the lead into ground water from the sites in which the collectors dump our trash.

Each of these features of natural social artifacts—their power, their scope, and their lack of any clear relation to intentional action or human design—needs to be considered at greater length. Let me begin with the last.

Artifacts and Intentions

The lack of any clear relation between a natural social artifact and intentions is important because it indicates just how difficult it can be to change such an artifact. Such an artifact has, as it were, a life of its own, a permanence, for instance, that resists change and a complexity that inhibits our knowledge of how to change it readily, without risking unforeseen and harmful effects.[11]

Consider the way in which we have acted to create clean and healthy drinking water. One of the landmarks in modern health care is the introduction in the late 1800s of municipal water plans, of citywide systems to provide all city residents with fresh and clean water. To provide such water to all residents, it is necessary to have a constant water supply and to have a system of arterial pipes to carry that water to the city, and it is convenient then to have additional pipes carry the water from those arteries of the water system to individual homes and to have pipes within each home so that water can be distributed wherever it is needed.

The end of all this planning seems straightforward enough, namely, to allow each person to turn on a tap in a home or apartment and get fresh water, uncontaminated by anything that might cause harm. But, clearly, each of the factors in producing clean and healthful water can be itself a source of problems. Problems can arise in the collection of water, in the waterworks themselves, in the delivery to homes and apartments, and in the homes and apartments themselves.[12]

For instance, guaranteeing the continued purity of the runoff into the lakes that supply a city's water may be no small problem, especially as more people move into the formerly pristine areas where the lakes tend to be located and as farmers make more and more use of fertilizers and insecticides.[13]

The recent problems with the parasite *Cryptosporidium* that plagued Milwaukee's water supply because of a "defective filtration system in one of the city's two water purification plants" is an indication of how difficult it can be to ensure both that water collected is pure[14] and that any impure water collected is made pure.[15] Water treatment plants are themselves complex mechanisms in which components are subject to breakdown or degradation. The difficulties of ensuring a safe supply are complicated by a parasite that is extremely small, "invisible to the naked eye," and difficult to test for,[16] and by inspection procedures that are, to put it mildly, less than adequate for a large percentage of water treatment plants in the United States.[17]

But the problems do not end with ensuring a pure water supply to the carrying system, for that system itself may introduce impurities. Current studies indicate that it introduces lead into the water we get from the tap,[18] and lead is harmful, especially to children.[19] The water in municipal water supplies is generally lead-free,[20] but the pipes that go to individual houses are often made of lead. Chicago's building code "actually required lead to be used until 1986."[21] In addition, the pipes that carry water through a house are connected with solder, and that solder was lead-based until 1988 when a change in its composition was mandated by law. Plumbing fixtures themselves can contain lead: even "new brass and bronze faucets can legally be as much as 8 percent lead by weight."[22] Anyone living in an older home is likely to have a potential problem, and the difficulty may be worse in newer homes because newer pipes are often a worse source of lead contamination as a result of not having "yet had a chance to acquire a protective layer."[23]

Our system of water distribution is an artifact, something we have created, and the complex web of interconnections we have is related in a relatively straightforward way to the intention to produce clean and healthy water for us all. What the lead in our water illustrates is how difficult it is, in acting with good intent to produce a good end, not to produce wholly unintended effects, an effect, in this case, directly contrary to the end such a system was designed to achieve. What those who designed such systems meant to achieve, presumably, is the assurance for each and every citizen within a particular water system that water from the tap is fresh and uncontaminated; yet the effect of the system designed to achieve that end is that "no household can assume its water is free of lead."[24]

Only someone with a conspiratorial cast of mind would suppose that the introduction of lead into our water supply is the result of some direct intention to cause harm.[25] Yet, as the example illustrates, it is hard to avoid bad effects even when there is an intent to produce good and where there is a relatively straightforward relation between that intention and the end produced to achieve it.

Such examples can be multiplied endlessly. For instance, farmers have discovered that a fungicide, Benlate, used to control molds and fungi to protect millions of dollars worth of crops so poisons the soil that "even weeds will not grow. Cucumber seeds will not germinate. Broccoli plants wither and die."[26] Farmers have acted with the end of protecting their crops but have so damaged the land as to make it impossible to grow the crops the fungicide was meant to protect. In trying to keep our crops from harm, we have made it impossible to grow them.[27]

There is no simple relationship between what we achieve and the intentions necessary to produce the results even where the intention is relatively clear—where the artifact constituted by our current water supply system, for instance, would not have existed but for the intention or set of intentions to produce such a supply. The situation is even more problematic for other natural social artifacts, many of which do not come into existence because of any clear intention to produce them. Not *designed* for any particular end or set of ends related to them, they have effects that may be far removed from whatever the intentions were of those whose acts and omissions effectively created them.

Consider our current system of land use and its effect on our rivers and streams—although to call it a *system*, or even a practice, may already be to categorize it in a way that belies how its bits and pieces have come into existence. People move to the country by a stream and cut trees and brush at the bank's edge to improve the view. The bank becomes more subject to erosion, which, when it occurs, widens the stream, allowing it to carry away more sediment and making it "shallower and warmer . . . unfit for many vital organisms."[28] It is not the home owners' intent to kill off any of the stream's inhabitants or so change its nature that less desirable fish find it significantly more congenial. Yet, that is what happens if enough people, acting out of aesthetic preferences, perhaps, open up their views of the stream.

Calling such a result a *practice* of land use, or an *artifact*, should not imply that all of its features or, indeed, any of its features are the result of intentional action. Nothing that happens to the stream—for instance its being widened, its carrying away more sediment as a consequence, its becoming shallower and thus warmer in the summer months when the sun beats down on it, and its therefore becoming a different sort of habitat, congenial to a different set of life forms than it formerly was—is the result of anyone's intention to change the stream.

Like the introduction of lead into our drinking water, the fundamental alteration of such a stream is an unintended effect of other intentions. In the case of lead in our water, the unintended effect is directly contrary to the original intent to produce uncontaminated water.[29] In the case of the stream, the unintended effects bear nothing but an unforeseen causal relation to the original intentions that produced them. In each case, we have a practice with a design—a pattern of land use, in the one, and a pattern of water use, in the other—that bears little relation to what any reasonably objective observer would design if given the opportunity. The quick and dirty test here is David Hume's: "Were a stranger to drop, in a sudden, into this world, I would show him, as a specimen of its ills, an hospital full of diseases, a prison crowded with malefactors and debtors, a field of battle strowed with carcases, a fleet floundering in the ocean, a nation languishing under tyranny, famine, or pestilence."[30] Hume uses the conceit to argue that we have no way of inferring from the nature of the world that it was designed by a God who is all-powerful, all-knowing, and all-good: If it were designed by some being, that being could not prevent evil from occurring, Hume argues, or did not know enough to know it would occur, or wanted it to occur or acquiesced in its occurrence.

We can use the same conceit of the objective reasonable stranger to assess our own practices,[31] and we will find that few, if any, bear any reasonable relation to what they would be were they designed for the ends they might be thought to serve. It is not just that a natural social artifact designed for clearly intended ends, such as municipal water supplies, have unintended consequences that are often counterproductive for the ends for which the system is designed but that an artifact itself is often not the result of any single intention or set of intentions but the creation of innumerable acts and omissions by individuals, corporations, or governmental officials that serves many, sometimes incompatible, ends.

Our land-use practices fall into the latter category, as the example of the effects of homeowners' actions regarding trees and brush on streams shows. We can add to that single example numerous other features of how we use our land that amount, all told, to a practice regarding its use that bears little, if any, relation to what a reasonably objective observer would design.

" 'If you take a drive out into pretty, rolling farm country, nobody thinks of the farming activity as habitat destruction,' says Dr. J. David Allan, a freshwater ecologist at the University of Michigan. 'But the transformation of the landscape by agriculture is taking its toll' on life in rivers and streams." Because it "is so much a part of deeply entrenched patterns of land and water use," Dr. Allan adds, "it is also far harder to deal with," even though it is "far more destructive to aquatic life," for instance, "than are spills of oil or toxic chemicals."[32]

No one would accuse farmers of acting with the intent to degrade the environment by, for example, polluting streams with the runoff from their fields or increasing the sediment in streams and so suffocating its aquatic life by the runoff of plowed dirt. But their acts and omissions produce a practice that has those effects.[33]

In the same way, we can use Hume's quick and dirty test to assess our practice of waste management.[34] We need to ask ourselves what any practice of waste management ought to accomplish. What ought to be its ends? And how well does our current practice achieve those ends, if at all?

These turn out to be contentious questions, as we will discover when we examine them in more detail in Chapters 3 and 4. But we can lay down some conditions for any practice of waste management that are relatively noncontentious. Here are two:

> 1. One essential condition is that the waste we dispose of not come back to us, either in the same form or in some more toxic form. That is, one condition we want to satisfy in any practice of waste management is to rid ourselves of the waste.
>
> 2. Another condition is that we not cause more harm by getting rid of something than would be caused by keeping it.

These seem modest enough ends. No doubt everyone has thrown something away and had it come right back. It is like throwing something out of the front window of a car and having it come in through a back window. And no doubt everyone has had, or heard of, the experience of causing more harm by throwing something away than would have been caused by keeping it. That would be like putting out medicine one no longer needed, or rat poison one suspected might be old, only to have someone or something get into the trash and cause the medication or the poison to be spread around where someone, a small child, perhaps, might inadvertently eat it.

Were a stranger "to drop, in a sudden, into this world" of waste disposal, he or she would be hard-pressed to determine what rational relations it bears to these modest ends of disposing. We act in such a way as to rid ourselves of the waste we produce in our homes only to find ourselves having it reenter our lives when, for instance, it leaks into the ground water that supplies our water from the sites to which waste haulers carry our refuse. Or we take our dogs for walks to ensure that they do not defecate in our homes and apartments, and their defecation enters the storm sewers and pollutes our rivers and sounds. Or we build incinerators to burn off our waste products and find that they emit toxic fumes we cannot help but breathe and leave concentrated residues of highly toxic compounds that are significantly more expensive to bury without worry about eventual harm to our water supplies, for instance.[35] Or we carefully act to ensure that what we throw away will be recycled, only to discover, despite all our trouble, that it is being treated as waste, not as a resource.[36]

Our system of waste management ought to have an end or set of ends for which it is designed, but the actual practice bears little relation to even the modest ends we ought to impose on any such system—that it get rid of what we dispose of and that it not cause more harm than would have been caused otherwise. Either the practice includes unintended effects that run counter to the ends that directed the creation of the practice, as with what has happened with lead in our municipal water supplies, or the practice grew without much regard to any ends having to do with what a rational observer might now take to be the point of the practice, as with what has happened to streams because of the way we use land along their banks.

Seeing the obvious effects on a stream of removing vegetation from its edges, a stranger dropped into this world might wonder if those effects were the point of the practice when the effects had never entered the minds of those whose ends were directed not by any concern about the river but by aesthetic considerations. Similarly, when we look at the ground-water pollution caused by badly sited and poorly contained municipal dumps, we might wonder in our worst moments whether producing such an end was the point of the practice as the effect achieved is so clearly now, after the fact, a predictable consequence of the practice.

But to take that cast of mind—of finding some evil intent behind the practice that causes such harm—is to suppose that for every practice with a design there is a designer. As I remarked in the Preface, Bernard Bailyn suggests that such a cast of mind is at least partially behind our Revolutionary War. We Americans were unable to comprehend how it was that a series of British acts—the Stamp Act, the sending of missionaries to Cambridge, Massachusetts, the Townshend Duties—could not be the result of "a deliberate assault launched surreptitiously by plotters against liberty both in England and in America."[37] If it was not the king's doing, it was the cabal behind the king. The British similarly saw the various acts of rebellion in such far-flung colonies as Massachusetts and Virginia as themselves parts of a master design, a conspiracy to throw off the English yoke. The difficulty both had was the difficulty of understanding how a practice, or sets of acts, with a pattern or design could have occurred without a designer.

Yet, we are all perfectly familiar with practices with designs that are not the result of anyone's intention. Darwin's theory of evolution is an instance of what have been called invisible-hand explanations.[38] Bats appear well-designed to catch insects without the use of sight, but their clever features are not the result of anyone's intentions, on Darwin's analysis, but the natural result of evolutionary processes in which the combination of some genetic traits with the environment insured the survival of certain features over others.[39] In short, it is easy enough to imagine something with a design coming about through the acts and omissions of individuals who had no particular intention to produce the particular design the combination of their acts and omissions with those of others has produced. The natives of Louisville say "Lou-ah-ville," and that has a pattern, but it is difficult to imagine that anyone, with knowledge and purpose, originally went about encouraging such a pronunciation.

So it is not unusual to have a natural social artifact—such as our way of disposing of our wastes—with a design that bears little relation to what any rational objective observer might create. Unfortunately, natural social artifacts can have a complexity to them and a permanence in their features that makes them difficult to dislodge or change even when, by use of such a test as Hume's, we begin to come to grips with them as practices that can be changed. What is worse, natural social artifacts also so affect our lives, because they are so pervasive and exercise such power over even our capacities to conceptualize alternatives, that it is difficult to get outside of them, as it were, to discern their limits and to assess them as practices.[40] Hume's quick and dirty test only works if it is applied, and it is not applied, sometimes, to those practices most in need of assessment.[41]

The Scope of Natural Artifacts

It is easy enough to think of natural social artifacts that are localized, are relatively easy to discern, and are open for assessment. The ways in which we handle dirty dishes is one example. The ways in which we take care of the leaves on our lawns, where we have trees and bushes, is another. But, in fact, such micro-natural artifacts, if I may call them that, tend to be part of larger natural artifacts that both create and reflect ways of looking at the world and our place in it and are, for that reason, harder to consider *as* artifacts.

Consider again what we might now call "our land-use practice." People purchase land, clear wood from vast parts of it, clear out what rocks they can that get in the way, lay out fields by creating boundaries, and begin to farm. In clearing the woods, they destroy the habitat of the animals and birds and the insect life that was there before; in burning the wood, they pollute the air; in clearing the land, they create earth less likely and less able to retain water so that the ground water will not be as readily replenished and the streams nearby will become more filled with silt, and so on. They grow crops on the land they have cleared, and they fertilize the ground to make the crops grow bigger, use insecticides to keep down the pests that would otherwise eat the crops, and lay down weed killers to encourage the growth of the crops. In doing such things, they are engaged in what is considered rational farming—not in habitat destruction, not in river polluting, not in decreasing the total number and kinds of living creatures.

The practice of farming in this way, which is widespread in many parts of the United States, combines with other practices—such as that of cutting trees and bushes along a bank to open the view or putting grass on the space around our houses and using pesticides to keep down the weeds—to create a network of practices that we may come to call "our land-use practice." Seeing that we have such a practice and that it is not inevitable but is the result of innumerable acts and omissions by any of a number of people is difficult, even though each part of the practice is itself a practice that may be readily discernible as a practice and open for assessment.

For instance, we in the United States tend to make the boundaries of our fields as open as we can whereas farmers in Great Britain have hedges. We can assess the different practices and see them *as* practices, as artifacts of different ways of using the land, but because they reflect different ways of using the land, we may be as blind to them as artifacts as we are to our very conception of how land is to be used.

Language is a clear example of how the scope of a natural social artifact can blind us to its being an artifact. The particular form a language takes, the particular vocabulary we use, the particular colloquialisms and modes of expression are the result of innumerable decisions made by individual users and picked up or discarded by others, consciously or not. I came to pronounce "milk" as I do at my parents' knees, without any choice, and the way I pronounce it is so habitual as to be almost impossible to change despite its rhyming with "elk" instead of "ilk" and so marking me in certain ways. It had not even occurred to me that I might be mispronouncing it until someone remarked on it, much too long into my life. Our forms of pronunciation are, as it were, micro-artifacts that serve us well, or well enough, and we are not likely to be aware of them as artifacts any more, for example, than we may be aware that the particular way we walk is an artifact and so can be changed—though only with a great deal of work and intensive training in a new way of carrying one's body and moving.

Even if we become aware of the features of a particular natural social artifact as an artifact, the network of micro-artifacts is so complex, with so many interconnections, that it is difficult for us to have any idea how to proceed to change any part of it. Understanding that a particular natural social artifact may be a complex of micro-artifacts gets us no purchase on how to alter the social artifact if that is our aim. It is too often unclear what is cause and what is effect and, as we have seen, too often unclear how to intervene in ways that do not produce untoward effects or even effects exactly the opposite of what we intend by intervening. Those who argued for pooper-scooper laws can only be frustrated by how little it has affected the practices of dog owners. How few carry scoopers is a symptom of how deeply embedded and difficult to change the practice is, or how inappropriate the means chosen to effectuate change, even when the unfortunate effects are known. For whatever reasons, "le weekend" is now widely used by the French despite the most persistent and best efforts of the French Academy to keep the language pure of English interlopers.

The anatomy of the success or failure of a particular word or phrase is an example of how a single utterance or use by a single individual can produce reverberations throughout the system of language if it has survival value. We might take seriously the Darwinian metaphor and take a particular natural social artifact, such as a natural language, as an entity, such as an animal, that has evolved into the particular form it has and is continually evolving under the particular relevant pressures. For animals, the pressures include climactic changes, the introduction of alien species into the environment, new diseases for which the animal has no immunity, and so on. For language, the pressures include new entities for which there is no vocabulary (e.g., the telephone, cellular phones, or CDs), the linguistic equivalent of climactic changes when one language clashes with another (as in some parts of the United States where Spanish is the dominant language and is affecting how English is spoken), and so on.[42]

Working out the details of such evolutionary behavior is intriguing and would no doubt be instructive.[43] Understanding, for instance, how the concept of privacy has changed since its introduction in the law in the late 1800s is a useful enterprise because it marks out how our understanding has changed in response to new ways in which we can be harmed,[44] and investigating the ways in which various objects have come to be regarded as waste would tell us much of the evolution of our concerns and forms of control over our environment. The same is true for how various forms of waste have come to be regarded as resources.[45]

But despite its changing over time in response to what appear to be evolutionary pressures,[46] a particular social artifact can so permeate our lives as to make acts or omissions seem inevitable in the way in which a particular word may come, unbidden, without conscious effort or thought, or in the way in which we crumple a can for the trash without thinking about alternatives. Natural social artifacts evolve, as any natural objects do, but they also function as normative systems, setting standards for those within them and so dominating the modes of thought as to squeeze out alternatives that might otherwise, but for the contingency of having this social artifact rather than that, rise to consciousness. Natural social artifacts have a normative power over both our thought and our actions. Because they tend to evolve only slowly and even sudden revolutions can quickly be assimilated within the ongoing system, those within them can readily be unaware of their contingency. Without a wrenching of our thought, it may not occur to us that they are artifacts, capable of being changed.

Normative Artifacts

When I wish to be understood, I speak in certain ways, and I risk misunderstanding or a complete failure to communicate by using an extravagant metaphor or newly coined word. Our natural language is a normative system: In order to communicate, one *must* speak or write in certain ways and *cannot* write in other ways.[47] There is leeway, of course. Our pronunciations can vary, as regional dialects show, and structure can strain the bounds of grammar and still be acceptable. Otherwise, we would never understand mistakes. Jokes about misspellings come off as jokes only because we understand how the mistaken spelling approaches accuracy and yet fails. But there are limits to what can be understood and thus limits to acceptability that, however difficult to demarcate practically or theoretically, cast some expressions out as complete failures.

A natural language has such a power over its speakers that it may be difficult for them to conceive of how to get out from under it. Those who are learning a foreign language can have difficulties learning to think in the foreign language, and that difficulty is a measure of how permanent a feature of our thoughts an obviously contingent social artifact can be. It is not just that we whose native tongue is English say, "It is raining," we *think* it as well and, without more ado, take it as a norm for how to say that it is raining. The forms of expression of our language are contingent both because we need not say things in just one way and because the language itself is contingent. Yet, like that Englishman who was happy to be back where people spoke the way they thought, we find the forms of expression we use to be the norm.[48]

Having been reared within a particular natural artifact, within a particular language, a particular mode of land use, or a particular way of disposing of our waste, we regard such matters from an internal point of view.[49] If we are not to be estranged in our ordinary lives, we simply live within those artifacts, without ever considering their contingency. But I risk certain sorts of misunderstanding if, while in England, for instance, I am as friendly and open as I am in the Midwest of the United States. The norms of how to respond to strangers differ, and not to alter one's norms to suit the altered circumstances is to risk a clash of norms—or at least misunderstandings. A German student at a college in the Midwest, who was in the United States for the first time, asked his advisor after several weeks whether something was *wrong* with him. The advisor was puzzled and asked why the student asked. He said, "The people on the street all look at me and smile!"

Someone who is external to a natural social artifact, a stranger dropped into our world, will see it as a contingent and perhaps puzzling feature of our lives—however pervasive and permanent it may seem to those who live within in. A stranger might look at the way we Americans stop at stop signs and come to a generalization or set of generalizations about our practices. When we are in motor vehicles and come to a sign of a certain shape, we either stop, look, and then proceed, or we slow down, look, and then proceed, executing, in the latter case, what some Americans call a California stop. That generalization states what those within the system do from the point of view of someone not in the system.

From the point of view of someone within the system, a stop sign is a sign at which one *ought* to stop. It is a norm within the system, accepted by those within, to stop, or at least slow down significantly, at such signs. It is not (just) a sign at which they will stop or slow down, as it is for the external observer, but a sign at which they ought to stop. It is not (just) predictive of behavior for someone within the system but determinative of it. Not stopping requires a justification.

The difference is the difference between being a member of a society in which there is an accepted rule about stopping at stop signs and being an observer of such a society, between taking an internal and an external point of view. In the latter case, one can observe regularities between such signs and behavior and state those regularities in generalizations about the behavior of the inhabitants. In the former case, one has internalized a rule: It is a norm of the system of which one is a member.

The same distinction holds for all natural social artifacts. We can observe how those within it behave and correlate various activities or omissions with other features of the system, or we can, from within the artifact, articulate its norms—what those who are its practitioners must (generally) do to be its practitioners. Someone trying to learn English will note correlations: Native speakers say "Hi!" in such and such contexts, "Bye!" in other contexts. But they need to know that to say "Bye!" when you meet someone has to be a joke or some sort of mistake, a Freudian slip, perhaps. You will have broken a norm and so risk misunderstanding or embarrassment. To learn the language is to come to take the internal point of view and so to know that saying "Bye!" when you meant "Hi!" is some sort of lapse, perhaps caused by your wishing you did not have to engage in conversation.

We need only string together a list of examples drawn from our various practices to get a sense of how internalized and thus compelling such practices can be. People who insist on recycling the waste they produce rather than hooking up to a city's sewer system are in for a long bout with the legal authorities and ultimate failure. Someone who insists on having natural childbirth in her own home without the help of any physician or other health care practitioner is presumed unreasonable, endangering herself and her future child by a desire that makes no good medical sense. Those who do not cut their grass may be thought to be ill or gone, and then, after a neighbor checks and discovers the omission to be deliberate, they may be thought to be obstinate and odd.

"Obstinate" and "odd" are terms of criticism,[50] and they mark one way in which those within a particular social artifact enforce its norms.[51] The norms primarily depend for their normative power on our internalizing them, rearing our children within them and criticizing even those who inadvertently or unintentionally break them.[52] But some norms reach the level of explicit recognition and sanction. The norm of hooking up the main water supply within a city to a house or apartment by using lead pipes was not just the norm for any plumber but was, as we have seen, required by law in Chicago until 1986. And some norms are enforced by making it virtually impossible not to follow them.

For instance, those who object to a society in which milk is purchased in boxes or plastic containers that are thrown away are unable to opt for anything else—even if the thought of other possibilities should occur to them. The system is so structured that other choices are rare. If we are to drink milk, we must purchase it as it is sold if we are not on a dairy farm. We have a way of handling such products that is not necessary: Our mode of packaging is an artifact that could be different, as we can discover by noting that milk used to be delivered in glass bottles to the door or by noting that in other countries one can purchase milk in plastic bags that can be recycled. But the artifact we have gives us no choice: We buy milk in plastic containers or cardboard cartons or not at all. Even if we become aware that the array of choices is itself an artifact, and not a necessary one, we cannot live outside its range.

Such internal regulation of choices is an exemplar of how a natural social artifact so dominates and permeates our lives as sometimes to exclude even realizing that there are alternatives, let alone trying to live outside the range of the artifact. This difficulty of what we might call the internalization of choices is especially acute for matters regarding environmental decisions because, I argue, the form of reasoning thought appropriate for such decisions—what is taken to count as reasonable and as unreasonable—is itself an artifact of our current practices and both extends them and is confirmed by them. To break with our current practices requires replacing our current model of how decisions ought to be made with a different decision-procedure, and the replacement is one that is not now a serious alternative within the community of those who are charged, in this society, with giving advice about how to make such decisions or with those who must make such decisions. Arguing for this new decision-procedure for those within our current natural artifact is like going to the grocery story and asking for milk in plastic pouches.

It is not unusual for a particular form of reasoning to be institutionalized within some ongoing practice. Accountants are deeply concerned not to have any item in their ledgers that has not been verified: they abhor figures that depend on predictions about costs, for instance, no matter how firmly grounded such predictions may be. An unverified item introduces an uncertainty into the system that will affect every calculation dependent on it. So all systems of accounting have a presumptive requirement that the items being added and deleted square up with *known* costs and debits—a requirement that can only be overcome by some showing of, for instance, great public need. The recent disputes about whether to include in the balance sheets of companies the stock options given executives hinges on the issue of whether the need for such information is worth what must be considered, from the accounting perspective, as the degradation of the accounting system.

One consequence of this requirement is that no one who reads a statement from a savings and loan institution can have any idea of its true worth—what money we would have in hand if we were to sell off all its assets and pay off all its debts. The reason is that accounting rules require that assets on the books be listed at their cost when they became assets.

So if a building was purchased in 1981 for $10 million, it remains worth $10 million as far as the books are concerned until it is sold and a new figure can be put in its place. But it may have been discovered after it was purchased that the building is on a polluted site and cannot be saved.

The loss of the original $10 million, plus the estimated costs of tearing the building down and cleaning up the site, may now make the building's value −$100 million. Yet, such a figure cannot be calculated into the total assets of the savings and loan until it comes to pass. *Estimated* costs that *may* be incurred do not count *as* costs. So the rules thought essential to proper accounting produce a figure that may bear little relation to the actual worth of a company.[53]

The accounting rules that necessitate such a result are thought essential to the way in which accountants reason about the worth of a company, and although we can understand the concern not to infect a balance sheet with what accountants would call guesses, the rules prevent even an accountant from knowing the true worth of such companies with a thorough look at the books and the competence to understand them. One result is the possibility of such debacles as the savings and loan crisis.[54] It is not just that inattentive investors, with the hope of great profits, gave money to unscrupulous managers, but that even the most well-informed investor, careful to read all the books in detail, could not have known that such investments were not wise or prudent.

In short, harm can come from such accounting rules, even though they may have been adopted with the best of intentions,[55] with quite reasonable justifications, and harm can come from any natural social artifact, whether intended or not. Our health care system cuts on a bias across lines that disproportionately harm women,[56] minorities,[57] youths,[58] and sick newborns[59] by denying them the same sort of treatment others in the system get. We need not argue that these exclusionary results are intended by anyone to raise the moral question whether we ought to have a health care system in which they occur. Similarly, if our practices of waste and resource use cause harms, then, whether intended or not, we have a moral issue: Ought we to have such an artifact?[60]

Morality and Values

As I have argued, a natural social artifact may have such a broad scope and its norms may be so internalized that we find it almost impossible to get out of its grip even though it may be oddly disconnected from any end or goal someone who might design such an artifact in response to some natural needs and wants would try to achieve. It is not just that it is difficult to get outside the scope of the artifact—such as speaking English without speaking English—but *wrong* to do so. Someone who pushes the envelope of English, as it were, risks not only misunderstanding but censure. As we have seen, a natural social artifact is itself a normative system.

Such artifacts no doubt reflect value choices as well as impose them. Spoken French is a language of elisions, and the result is mellifluous. Whether it was the ease of pronunciation, the pleasure of the sound, or something else that produced the elisions, the effect reflects such value choices as that ease of pronunciation is more important because more efficient than a strict adherence to the sounds of every particular nook and cranny of the written words or that the flow of sound produced by elision is more important, because it sounds better, than the staccato effect produced by a careful pronunciation of every word as distinct. Just as we will not be understood or, if understood, censured, if we do not elide where we ought to when we try to speak French, so the choices speakers of the language make will in turn reflect the values that they, consciously or not, impose on their use of the language. The use of "le weekend" is not a conspiracy by French anglophiles to subvert the French language, but the result of countless choices by French speakers, who no doubt often have nothing more in mind than communicating, clearly, their plans for the weekend.

Likewise, our natural social artifact of waste management reflects value choices—to put landfills out of sight and out of the range of smell of most people, to manufacture goods with the least expense so that the refuse of manufacturing is disposed of as cheaply as possible, and so on. And just as it reflects value choices, it imposes them. As we have seen, purchasing milk in glass or plastic pouches is difficult, if not impossible, for most in the United States. Those who value recycling of containers have no choice.

Different forms of an artifact no doubt reflect and impose different sets of values and have differing moral implications.[61] For instance, a judicial system committed to trial by jury presumes that ordinary citizens are competent to make sometimes quite complicated judgments of guilt or innocence or that, at the least, it is better to make that presumption than any other. A system of administrative trial presumes what we might call an arrogance of judgment on the part of whoever is determining guilt or innocence. An administrative hearing, in which the person to whom accusations have been brought is examining the evidence, hearing the accused, and assessing guilt or innocence, presumes that the administrator is somehow more worthy of making judgment of guilt and innocence and less likely to make mistakes than fellow citizens. Choosing a judicial system in which administrators decide guilt or innocence certainly reflects value choices. In accusations of sexual harassment, an administrator may be trying to protect the privacy of those bringing the accusations by having administrative hearings, for instance, but the choice of differing judicial procedures has different moral implications. In trial by jury, the judge is presumed to have, by the structure of the procedure, no more moral insight into guilt and innocence than the proverbial person on the street; in administrative hearings, the administrator must be presumed to have an elevated moral status for the hearings to be morally defensible.

Differing values are also reflected in different ways of responding to waste, and different moral implications flow from different forms of waste management. Underlying our ongoing natural social artifact of waste management is a particular mode of reasoning that both helps produce the artifact and further entrenches it. In reflecting a vision of how rational decisions are made, that mode of reasoning reflects certain values, but to the extent that that mode helps produce a natural social artifact that causes harm, we make a mistake in using it. If we can change our current artifact for the better, with less harmful results, we have a moral obligation to do so. But to see how to do so, by changing the way we make decisions about waste, we need to consider the way in which we come to decisions regarding waste.

CHAPTER THREE

Knowledge and Doubt

A natural social artifact, I have suggested, may contain as an essential part a particular form of decision making, a way of thinking about issues that makes alternative forms of decision making seem inappropriate, if not irrational. Someone who takes an occasional flier on race horses has a sense of how different the cast of mind must be for someone who approaches life in general, in all its aspects, as a flier, but there can be deep difficulties in coming to grips with a different cast of mind, a different decision-procedure. The problem is that adopting a different decision-procedure may seem not just like a different way to approach the world—as a game, say, more than as something more serious—but as a sign of something thoroughly wrong. One gets a sense of how deep the difference can go in discussions about where to site incineration plants[1] or where to put low-level radiation dumps[2] when the proponents argue that the known risk is not especially high and opponents retort that any risk is too high. The two sides seem to bypass each other, and part of what is happening, I suggest, is that they are appealing to different models about how we ought to make rational decisions about such matters.

Their risk assessments and their perceptions of what risks they face differ because of competing models with differing conceptions of how to take risk into account.[3] It will not help to attempt to assess the various risk assessments unless we first back off and examine the competing models.[4] For the criteria for assessing are skewed by the different models each party thinks appropriate.

These disputes are like disputes about abortion, for instance, where the opposing sides seem to take such different stances about the matter that no common ground is left. The pathology of such disputes is a matter of some importance given how they can cause such difficulties to the body politic. I do not know whether the sort of claims I make here about competing models of decision making and about how to choose between them can be generalized to such other similar sorts of disputes, but the dispute certainly has some of the features of our having an essentially contestable ideal of rational decision making, an ideal supporting competing conceptions of how we ought to make reasonable decisions.

It is common in such disputes to find that each party finds the other inconceivably dense, if I may put it that way, unable to comprehend what seems obvious. As I emphasized in Chapter 1, it is hard for those who think we need all relevant information to make public policy decisions to understand how someone could seriously think we should act in ignorance or, worse, that sometimes we *ought* to act in ignorance—and can do so reasonably.[5]

It is also common to find that these competing conceptions are each held up as paradigms of how to proceed, delaying movement even when they are not themselves held by those in positions of power. This is especially so because it is always appropriate to ask for further information, and that demand may readily be taken as a demand that must be satisfied before any public policy decision can be made rather than simply as a demand that propels further research. In short, if we are not clear exactly *why* we are making decisions under one conception rather than the other, we may find ourselves waylaid by what may seem perfectly appropriate appeals that will exert a gravitational pull on our decisions, delaying them or forcing us to modify them in light of the competing conception.

This can happen especially in an area of great dispute, where matters are unsettled, so that sometimes decisions are made with one conception in mind and sometimes with another. The latter pattern is typical of what has been happening regarding environmental issues in the 1980s. So when I claim that we need to displace the one form of decision-procedure with another, I mean that we need to give it priority and that we need to be clear *why* it should have priority—what reasons make it moral and reasonable to use it. I do not mean to suggest that the decision-procedure has not been used, but that without clarity about why it ought to be used, we shall find its use sporadic, lacking the unity of purpose that will give it the continuity it needs if we are to make a run at recovering even a portion of Eden.

The sort of conflict that results when one form of decision-procedure does not have priority of place can be illustrated in the conflict over whether the rate of extinction of wildlife is increasing or has remained relatively constant. Some claim that, as the World Wildlife Fund is quoted as putting it, "Without firing a shot, we may kill one-fifth of all species of life on this planet in the next 10 years." Others claim that there is no evidence that the rate of extinction is going up.[6] We may think the first an exaggeration without being at all convinced that the second is really quite relevant. These claims seem to run right past each other, in any event, for what is at issue is what sort of projections we ought to make given the potential loss of species. With forests being cut at greater and greater speeds, for instance, are we to presume that the rate will increase or not? Past evidence of depletion of forest cover without much loss of bird life, for instance, does not seem decisive as a basis for projection because so much more is being cut in what appear to be habitats richer in varieties and numbers of life.[7] So what is at issue here in deciding what our public policy ought to be regarding such potential loss is what *presumption* is the reasonable one to make. On this view of the matter, both presumptions serve the same logical function in any arguments about the rate of extinction, and presuming no increase in the rate of extinction is thus as much in need of support *as* a presumption as presuming a rate of extinction that would guarantee the loss of one-fifth of all species of life.

There is an additional problem we should remind ourselves of. The way we now think about using our natural resources is incoherent. We introduce a new chemical to solve some special problem, or fasten on a new manufacturing technique that is more efficient and less costly, or build a new plant better designed to serve its manufacturing ends, or purchase a lot along a river and cut down the trees and vegetation so we can see the water. A theoretical justification for such laissez-faire decision making can be provided, turning on the freedom of individuals and corporations to exploit resources for their own benefit and/or for the common good. Such decisions are made without full information, and no one in the midst of our current natural social artifact thinks them any the worse for that. Yet once our view of the river is secured, once a plant is in place, once a chemical is in the environment, those who think harm is being done must respond to the challenge, "But how can you be sure?" Full information is *then* required.

Thus, one curious paradox of our current way of using the earth's resources is that we make use of two different models of how to make rational decisions about such use. We allow use and exploitation without demanding full information and then demand full information to prevent such use. The paradigm that requires full information is only invoked after decisions about how to use the earth's resources have been made on other grounds—such as the freedom of individuals to do what they wish with their own land.

Getting coherence in our view of how we ought to make decisions about the environment will be no easy task. I argue in Chapter 6 for a fundamental shift in the way in which we view our risks, especially when the magnitude of the potential harm is great. I begin here by laying out in more detail the sense that we always have an epistemic shortfall in public policy decisions and in decisions about the environment and waste in particular. This sense of a shortfall is a crucial component of the conception of decision making I mean to displace. I then argue that we often, indeed commonly, make reasonable decisions without all relevant information, and I lay out what is common to the conception of decision making I mean to displace and the one I propose. I lay out, that is, the situation we find ourselves in, and particularly the epistemic situation we are in, independently of what conception we end up adopting about how to make decisions about our situation.

The Epistemological Card

We have presumably all had occasions in which not having all the relevant information has caused us difficulties in making decisions. We have all had occasions where an ordinary mechanical or electrical device has caused us difficulty because of some failure on our part to comprehend fully how it works. So we have failed to awaken in time because the alarm is set for P.M. instead of A.M., or we have locked ourselves out of the house because we thought the new lock, which worked when the door was open, would certainly work when the door was shut.

It does not take many such occasions to make us cautious in approaching the world. There are few things more frustrating than to put in a glistening new lock for a door, test it several times while the door is open, close and lock the door, and then be locked out. In such situations, one wishes for godlike omniscience but recognizes that mistakes are going to be made, tries to minimize them by obtaining as much relevant information as one can (by reading carefully the instructions about installing locks), and protects oneself against the inevitable failures (by making sure one has the back door key before locking the front door).

The wish for godlike omniscience is, of course, an expression of what we would ideally want. Not being gods, we shall never achieve it, and being human, we would never know even if we were to achieve it. This epistemological shortfall is one source of a pervasive feature of public policy discussions, like those regarding waste. For it is always the case, first, that we will not know we have all the relevant information we need and, second, that more relevant facts may remain to be discovered. These are separate facts—one about us and one about the world—that conspire to a lack of closure in regard to many public policy matters.

Many practical problems have crisp ends—the proper spelling of a word, a simple math problem, and so on. We may have difficulty balancing our checkbooks, but there is, we know, an answer to the question how much money is, or is not, in an account. The complexities of the phenomena under discussion in public policy matters, however, always leave open the likelihood of there being more relevant facts to be uncovered—from the permeability or impermeability of various sorts of soils and clays that may line refuse dumps to the effects of microwave transmissions on genes to such mundane matters as the setting of dumping fees to encourage recycling and yet make sure a landfill is profitable.[8] And we are always in the position of knowing that such a likelihood exists because, we know, we are not omniscient and, besides, we have a great deal of evidence that after acting in the past, we have discovered new information that has thwarted our formerly reasonable projections and expectations.

Such mistakes in projections are not confined to environmental issues, obviously. They seem endemic to any major public construction project, for instance, and are no doubt an essential part of the structure of political dialogue that must accompany any proposal for such a project. Projected costs are low and projected benefits are high in order to encourage funding, and it is no surprise that there are cost overruns or that the benefits do not materialize. The original projections are always presented as reasonable, firmly based on evidence about costs and surveys of intended use.[9]

For instance, a permit was issued for an incinerator in Detroit based on a health risk assessment that showed an increased incidence of cancer of one in a million.[10] The next year an engineering calculation error was discovered that had caused the risk to be underestimated by a factor of 36. We did not know what we thought we knew, and information about the engineering calculations, which we would have thought irrelevant, turned out to be relevant. We always may lack relevant information, and, besides, the usual case is one in which we know we lack relevant information.

For example, modern techniques for removing heavy metals from emissions have encountered serious problems with mercury, which is highly toxic. It seems perfectly reasonable to resist installing expensive dry scrubber and baghouse equipment, which removes some heavy metals from emissions, when we know that such equipment has such failings and that more investigation may solve the problem.[11] If we do install such equipment, when it does not succeed in completely removing toxic wastes, we pay a heavy price. For instance, the huge incinerator in Detroit was shut down less than a year after it started because it was emitting "much higher levels of mercury than are permissible under state law."[12] The city was required to refit its boilers with scrubbers and more sophisticated pollution control devices if necessary, at a cost of $30 million through 1997. This expensive failure was caused, it is argued, by deciding to act without complete information.

So the demand for more relevant information is always appropriate in public policy matters. During the PBB crisis in Michigan, for instance, it was always appropriate to ask for more information about the effects of PBB on the health of those who had ingested it, and, I should add, the question remains appropriate. The request for a further study is a card that can always be played, for if any decision is to be made about a public policy issue—to build a dam in an environmentally sensitive area, or a landfill, or a nuclear power station—that decision will be made in the midst of what can always be an ongoing investigatory process. After I have balanced my checkbook, I may want to check my data—my checks and other debits, for instance—and doublecheck my calculations, but nothing about the normal situation of balancing one's checkbook makes it appropriate, on epistemological grounds, to ask for *more* information after one has doublechecked. The investigatory process has a point of closure when it is appropriate to write a check or not and inappropriate to insist, "But are you *sure?*" Public policy issues, however, tend not to have such clearly marked, crisp endings.

By the nature of the case, public policy issues present us with an epistemological shortfall: The scientific situations are too complex, and our own ignorance is too vast, for us ever to be able to reject as inappropriate on epistemological grounds the request for more information. We may want to reject such a request on other grounds. We may think that the risk of not acting is too great or the danger in acting so small that despite our not knowing that we have all relevant information, we should act anyway. But it is always appropriate to ask, "Are you sure the risk is too great?" or "Are you sure the danger of acting is so small?" The epistemological card can *always* be played in public policy matters whose resolution depends on complex scientific issues, and, unlike the case where one has doublechecked one's checkbook, it seems not unreasonable to play it. In addition, two other systemic features of public policy issues conspire to reinforce the lack of closure the epistemological shortfall encourages.

First, it is politically advantageous to refuse to act when it is epistemologically appropriate to ask for more information.[13] It always *looks* reasonable to refuse to act on the ground that "not all the information is in." Decision making is then delayed. An elected official can thus keep from taking any decisive action that might harm constituents' interests while claiming to be acting rationally[14] and can justifiably turn the decision making over to experts so that, when a decision is made, they can be blamed, by the constituents and perhaps the elected official alike.[15]

Second, the price of change is always high. We usually have a vested interest in maintaining the status quo: change creates uncertainties, nullifies past investments, guarantees future expenses we otherwise would not have, and so on. So we can generally justify change only with a preponderance of evidence in favor of it, and as long as an investigatory process *can* continue because we may not have all the relevant information, it seems always appropriate to ask for more information before changing our ways. We know from past experience that we have changed our ways and later discovered that we have not gained enough to justify the change. "It is better to be safe than sorry."

What we may call the epistemological card is thus powered by a confluence of concerns each in itself sufficient to power the demand. The stakes of change are too high, the scientific situation is too complex, and our own ignorance is too vast, it is claimed, to justify much self-assurance about stopping at some point in the process of deciding what to do about a public policy matter and saying, "Now we simply must decide what to do." Besides, our always being able to ask for further relevant information has the good effect of driving research, as in the case of emission scrubbers not fully capturing mercury.

So it is no surprise how disadvantaged someone is, in any discussion, to claim that sometimes, on some occasions, we should make decisions without waiting for further information that, all would agree, may be relevant to making a fully informed decision. The mere suggestion is treated as evidence that one is simply not rational, and the persons with the presumption find it hard to know how to proceed. "Is this person seriously suggesting," one sees them thinking, "that we should act *without* knowing that we know all we need to know in order to know we are acting correctly?"

Deciding Reasonably without All Relevant Information

Three responses need to be made. First, we often make reasonable decisions without all relevant information. Indeed, except for such restricted cases as simple mathematical calculations, it is difficult to imagine cases where we do act knowing that we have all relevant information. Second, if we wait for all the relevant information in regard to waste, we shall confound our own interests, for our inactivity has consequences—waste accumulates—and what we do with it while we wait creates practices that must be considered if we come to have more information to decide what to do.

Third, the goal is illusory, at least in regard to the sorts of problems we are considering, for we do not have any chance at all of obtaining all relevant information about waste—the effects of compounds when incinerated; the actual costs of building, operating, and maintaining incineration plants, including the training and related costs of those who are to run and manage them; the long-term effects of landfilling to ground and water contamination; and so on.

I wish to consider each of these responses in turn, but the aim is not just to remove the sting of the epistemological card, the presumption that it is always appropriate to ask for more relevant information before coming to a decision. What is objectionable is not that we should always seek more relevant information, but the presumption that we ought not to act until we have it all. The request for more information is reiterable, and although it might be reasonable to get some specific information before acting, and it is always appropriate to seek more information, the presumption that it is always appropriate to ask for more before acting is an impediment to appropriate decision making in regard to many environmental issues and to waste management in particular, but it is a compelling presumption, one not likely to disappear unless a replacement is in place that accommodates the quite appropriate demand for more relevant information and properly fits the situation. So a positive aim is to make clear what kind of situation we are in regarding waste and what sort of reasoning is appropriate to the problem. Let me begin with situations where we decide reasonably without all relevant information.

I risk some misunderstanding immediately because I begin with a kind of case that may be thought by some to confirm rather than undermine the presumption, namely, gambling. It is often thought that nothing could be more unreasonable than to gamble, but rules of reason operate in games of chance as well as elsewhere.

The crucial feature of gambling that is relevant here is that the conditions of play are so constructed that one cannot have all relevant knowledge. If one knew what the lotto number was going to be, one would choose it, the clear winner, and not some other number. If one knew the card the blackjack dealer was to deal, one would know whether to take or reject it.

Games of chance are not all of a kind, and if we were to work out the rules for what is reasonable in poker, for instance, we would find them not everywhere the same with what is reasonable in blackjack or playing the numbers. There are occasions when it is reasonable to bluff in poker and occasions when it is not. The occasions for bluffing in blackjack are fewer, and the style in which we bluff differs. Yet, bluffing has no place in playing the numbers: that is not in any way a game of skill but depends entirely on chance.

It is the power and extent of this feature of chance in gambling that so impresses those who think it is not rational. To gamble is to take a risk where luck, or bad luck, can completely determine the outcome, and the greater the play for chance, the greater the risk of luck playing the crucial role.

Yet, there are reasonable and unreasonable responses to such an eventuality. It is generally unreasonable to assume that one shall always win and that one should thus always bet so as to maximize one's possible benefits. People who act in this way are high fliers and rightly criticized for always choosing the 100-to-1 long shot. Such unjaded optimism in the face of a world of risk approaches irrationality. We follow what decision theorists call the maximax rule if we act with such optimism, assuming that "no matter which act we choose the best outcome compatible with it will eventuate."[16] Such a cavalier attitude is not inappropriate if we are betting with play money or nothing hangs on whether, or how much, we win or lose, and we know from such stories as Jack-in-the-beanstalk that high fliers are sometimes lucky. The story is a classic reminder that we can sometimes do something as apparently irrational as trade the family cow for a bean and everything will still end happily.

Yet, clearly more caution is usually more reasonable, and, in one sort of case at least, extreme caution seems the most reasonable. In fact, in some cases we ought to assume the worst and act like a deep pessimist. Suppose that among a number of options some have such bad consequences as to be totally unacceptable—death, financial ruin, and so on. Suppose also that no great gains attach to any of the other alternatives. If we cannot determine which alternative is most likely to occur, we ought to act as though that alternative with the least bad outcome will occur. We then follow what decision theorists call the maximin rule. "Assuming that whatever one does, the worst will happen, the maximin rules pick the best of the worst."[17] Given the price of failure, we choose to minimize our possible losses rather than, like a high flier, attempt to maximize our potential gains.

The attitude of mind captured by this betting strategy is the quite reasonable one of extreme caution where, among other things, the possible losses are unacceptably high. The careful placing of one's feet and hands on a sheer rock wall so as to minimize the harm if the rock gives way is similarly contoured to the salient facts of the situation. The care is dictated by a reasonable desire not to end up a high flier.

That we appreciate the need for care and caution in rock climbing, that we marvel at Jack's good fortune and hope our children and political leaders do not imitate him, that we can see how unreasonable it would be sometimes to bluff in poker—these and countless other examples indicate how, even if we were in situations artificially restricted to deny us all relevant knowledge, we can pick and choose among alternatives in reasonable ways. To put the point less optimistically, and in more general terms, some choices are, in some situations, more reasonable than others, even without all relevant information.

This general truth applies to a great many more situations than those in which we gamble. When we play checkers, for example, we need not gamble. The pieces are distributed in exactly the same way for both players, and nothing in the rules except, perhaps, who should begin depends on any element of chance. No dice are tossed, no cards drawn. We know what moves our opponent may make. The rules prohibit certain possibilities and permit others. In addition, we can make reasonable judgments about what moves our opponent will make. At some points, especially toward the end of the game, only one move may be possible or, if several are, only one move may be reasonable given a desire to win. At other points, too many possibilities may emerge, and we must judge, from our perspective, what our opponent is more or less likely to do. Because we make moves in games such as checkers in part in anticipation of our opponent's response, then, because we often do not know what our opponent will do, we decide, without all relevant information, on the basis of our best judgment of what our opponent is *likely* to do.[18]

In making such judgments, we may well lack information that would be helpful. Is our opponent a novice at the game or very experienced? Willing to take chances or exceedingly cautious? An open book or able to hide emotions of joy or worry that might give us a clue? Lacking such information, we do not thereby think that *any* move we make is equally reasonable. We assess our moves—and if we do not, any kibitzer will—on the basis of whether they will advance us toward our goal and take account of the probable moves of our opponent, given what we know is permitted, what is likely by someone fully experienced, keen to win, and on the game, and how near we judge our opponent to approach that ideal. Moves are judged as more or less reasonable depending on how they contour themselves to the salient facts available. We do not think them inherently irrational if they are made without all relevant information provided they take account of the information available.[19]

A great deal of ordinary life is like a game in that respect. When we drive a car, we know what moves are permitted ourselves and other drivers, and we quickly note in any situation if others are doing what is not permitted and are likely to continue doing so. Even the moves we make in passing another car, for instance, take account of the probable moves of other drivers and are reasonable, or unreasonable, even without all relevant knowledge of what others will do.

The comparison of much of life with a game, in which our knowledge is artificially restricted, should not mislead. I am not claiming that because in practical affairs all we ever have are probabilities—or, a weaker claim, because probabilities affect all actions we take—we cannot ever act with all relevant knowledge. That may be true, but putting the matter that way makes it sound as though it were a source of regret and of deep skepticism about empirical claims. The claim is rather that life goes on, whether we have all relevant information or not, and that because we act and omit acting in our lives as they go on, we often, indeed usually, act without all relevant knowledge and think, quite rightly, that we can act more or less reasonably. The source of concern is not general skepticism but realism about our roles as agents in a changing world.

When we consider examples of making decisions, we tend to isolate the situation, to concentrate on whether it is reasonable to move *this* piece given the locations of the other pieces, or whether we really ought to try to pass another vehicle in *this* situation. But other traffic proceeds whether one passes or not, and the game clock ticks away. Decisions cannot be isolated in that way. We are involved in *processes*, and those continue despite our attempting to isolate actions. We grow older while trying to decide how to look younger, and, in general, we almost always are in a situation where we do not have all relevant information, perhaps could get it, but not anytime soon, and need to act.

If we do not know whether a particular plant needs fertilizer, or is getting too little sun, or is too close to a walnut's poisonous roots for long-term survival, and we do not act because we lack all relevant knowledge, then the plant will respond to its conditions, and not wait for us to become fully knowledgeable. By the time we discover that it is too close to the walnut roots, it will be too late for the plant.[20] If one is a physician with a cancerous patient, and the cancer is spreading in ways unusual for a cancer of that sort, then clearly it would be best to know its rate of spread, its pattern and mechanism of dispersal, and other relevant features, and one should try to discover as much as one can as quickly as one can. But, in fact, one will have to treat without all the relevant information, basing one's treatment on how seemingly similar cancers have been treated, leaving open, if one can, room for other possibilities. For the patient will surely die without treatment: life and death go on without all relevant information.

It may seem that a physician in such a situation really has three possibilities, to treat, not to treat, and not to decide whether to treat or not. One may compare the physician's situation to that of someone trying to decide to purchase a new car: one may decide to purchase it, decide not to purchase it, or not decide one way or the other right now, waiting, perhaps, for more money or information. But the physician does not have that third option. The new car will still be there, we may presume, unchanged after one has hemmed and hawed. The patient's progress, or lack of progress, will continue unabated whether one decides not to treat or decides not to decide whether to treat. Those two options have the same effect on the patient, and given a physician's obligation to help a person taken on as a patient, he or she is responsible for the effects both of *no* treatment and of *non*treatment. The former may be appropriate in some diseases: flu is not effectively treatable, although some symptoms can be alleviated. But a doctor who failed to treat a patient when that would be the standard medical practice would not be excused if he or she were to say, "I didn't decide whether to treat, and that's not to *do* anything. That's just *not* deciding to do anything."[21]

We can generate examples here without difficulty. We may despair, for instance, of making reasonable choices about how to eat. We may think, "Some say meat is bad for you; some say some meat is good. They're always discovering that one shouldn't eat this or that after telling us we should eat that or this.[22] I can't keep track. They don't *know*. So I'll just eat what I like." But one choice we do not have is that of not deciding whether to eat. "Nothing goes in!" would be an irrational principle to adopt even in the way of little information about what is good for us: It is not a real possibility. Just as a physician cannot have as a real possibility of treatment for a cancerous patient not to decide whether to treat, so we cannot have as a real possibility not to decide whether to eat or not, provided we want to live. We will eat whether we do so reasonably or unreasonably, and how we eat, as we wait for all relevant information about how best to eat, will affect us, perhaps increasing or decreasing, despite our ignorance, the likelihood of getting cancer.

We can modify our eating habits as we gather more information, forsaking butter and then margarine, for instance, each time presuming that our choices are better for our having taken the additional information into account, but each time having to make a new decision without all relevant information. We might suppose that each new decision will make our eating more and more rational, because more and more informed— as though we were zeroing in on the best of all possible ways of eating. But unable to discern what is truly false and misleading from what is true, we can at best modify our habits as new claims are made, hoping for more than short-term gains in a relatively long life. Changing our eating habits sometimes requires enormous changes and sometimes produces significant results, but we still must do that without all relevant information. Even if the claims are correct, that is, we will still have an epistemic shortfall both because we are not in a position now to know that they are correct and because we do know that we are not now likely to have all relevant information about how to eat well. Deciding not to eat and deciding not to decide whether to eat in the face of such uncertainties are not different options—if we wish to live.

So it is with waste. Just as we eat, so we make waste. There are leftovers to be disposed of. At a minimum we defecate and urinate, and we will do so, if we eat and drink, quite independently of whether we wait for all relevant information about the best means of disposing of these things. By-products are the inevitable consequence of living, from our perspiration and the oil on our hands that are the bane of art curators around the world to the residue of whatever we use to warm ourselves and our homes. We can decide to do something about it or decide not to do anything about it, but it is not a real option not to decide what to do. The effects of not deciding are the same as deciding not to do anything, and things will change if we do not decide what to do. The waste will simply accumulate. We have, in short, no option but to decide without all relevant information.

If we do not know, for instance, what change in scrubbers, if any, will be completely effective in removing mercury from emissions, then, because we have no clear idea how long it will be before we have such information, if ever, we shall continue to have emissions with mercury and other toxic metals and compounds so long as we fail to act because we are waiting for more information. For instance, we have evidence that two large families of chlorinated organic compounds are formed *after* combustion—the polychlorinated dibenzodioxins (PCDDs) and the polychlorinated dibenzofurans (PCDFs).[23] They may be "products of the combustion process itself or of postcombustion chemical reactions,"[24] but the matter is a subject of much research and little agreement. Yet, they are highly toxic, and because they accumulate in fatty tissues, they are more concentrated the higher up the food chain you go. Information about their formation is thus important. But to wait to decide what to do about emission controls until we obtain sufficient agreement to justify action means that we will have decided not to do anything about the emissions that would continue unabated.

Such examples can be multiplied about almost every factor that can make a difference in deciding what to do about waste, but the point is a general one about the very nature of the problem we face: We have to decide without all relevant information.

The Situation We Are In

We think that the more accurately we can predict, the better the state of our knowledge. It is this assumption that underlies our feeling that we know our friends well, for we can often predict with a great deal of accuracy what they will do. But if we use this criterion of predictability to assess our knowledge of environmental dangers, then we lack knowledge, for we are unable to make certain predictions regarding many matters and often unable to make any proper predictions at all.[25]

The effects of the ingestion of polybrominated biphenyl (PBB) provide a good example. Polybrominated biphenyl is used, among other things, as a fire retardant in movie projectors, as insulation around the hot lights. So when bags of it were accidentally mixed with animal feed at a Farm Bureau lot in Michigan and contaminated animals began to die, everyone was at a loss initially to discover the cause. Who would have thought it would ever enter the food chain? No research existed to give a clue. Had scientists been asked to predict the effects of PBB's ingestion by humans, they could at best have extrapolated from similar compounds with presumedly similar effects, but they would not have known its effects with any certainty.[26]

Yet PBB is just one of thousands of chemical compounds to have been created in this country alone in the last half century. One may add here information about projected costs and subsequent cost overruns of waste disposal plants, or about any of the other variables about waste, but the point is that the situation we find ourselves in regarding waste, as with many environmental issues, has at least the following characteristics: (1) We do not know everything that may be relevant; and (2) whatever the state of our knowledge, we are not in a position not to do something.

It is the combination of both factors, added to the presumption that a fully rational decision needs all relevant information, that leads to the paradox of incoherence regarding our present practices remarked on in Chapter 1. As we put off deciding because we do *not* have all relevant information, we must still dispose of our waste, for instance, and so we do—filling landfills, incinerating trash and garbage, pumping brine into deep wells. Yet, we do these things without all relevant information, and so they are, given the presumption, less than fully rational.

I do not want to deny that these decisions may be less than fully rational. I do want to deny that we should accept a model of decision making that guarantees that we shall not make rational decisions, by its standards, even though we must continue to make decisions. The natural tendency of imposing such a standard is to make us less than fully careful and cautious about those decisions we make in the interim, to fill the gap as we try to get all relevant knowledge. It would be far better to determine how to make the most rational decisions given our ignorance.

It would not so much matter if our acts and omissions, made without full knowledge, did not have any consequences, but they do. For another pair of characteristics of the situation we are in regarding environmental issues and waste in particular is that (1) our present acts and omissions may, for all we can know, have a profoundly detrimental impact on our health and our environment; and (2) our present acts and omissions—

and the practices they engender—create precedents and states of affairs that are not easily changed.

Once we have built landfills, for instance, or have constructed incineration plants, there is an imperative to use them unless we can meet a very heavy burden of showing that they ought not be used. Such a justification would not just take a showing of great harm from their use but also require an attractive alternative at hand, one that, it must be shown, will not be discovered someday to surprise us with the same sort of adverse consequences as, say, the landfill or the incineration plant being replaced.[27] Given such requirements, present decisions have a built-in dead weight.

We have the same difficulty even when we do not make full-blown public policy decisions but simply let things go—by allowing leaf burning even when it is prohibited, by not regulating the carrying and thus the dumping of toxic wastes and so allowing clandestine dumping on back roads, and so on. These practices come to have a weight of their own and are a real claim on public policymakers. It is not an easy matter to assess this weight, but we all know how difficult it is to change our habits or a long-standing practice—to have citizens, for example, put their leaves at the curbs for pickup instead of burning them as they may have done for years, or to encourage citizens to recycle leaves as compost once they have become used to placing them at the curb.[28]

Yet, the dead weight of practices comes not just from the difficulty of changing deeply ingrained habits or of motivating anyone to change when changing a practice requires that most change and those who change first are out of step. Present practice has a *moral* weight as well. If, for example, a county commission decides in a comprehensive long-range plan that the primary county landfill will be in one area of the county, that decision frees the rest of the county from concern that their property values and quality of life will be affected by any landfill any time soon. Among other things, people will purchase new houses, renovate their old homes, build businesses—creating in the process substantial interests in maintaining those homes and businesses free of the smell and noise and unsightliness of a landfill. If the county commission reverses its decision four or five years later and puts a landfill in another area of the county, then some of those people who relied on that original decision will have some substantial interests harmed.

It is always morally wrong to harm the interests of others without good reason, and it is morally worse first to encourage people to cultivate such interests and then to harm them. That is the moral equivalent, in public policy matters, of enticing an individual to do a wrong—"Go ahead; it's O.K."—and then calling them out—"Mommy, Mommy, look what John did!" The "enticement" in public policy matters needs scare quotes because it is presumably not intentional, but the harm is substantially the same. It is prima facie morally wrong to take back what one has said when others can reasonably be expected to rely on what has been said.[29]

To say that it is prima facie wrong does not mean that we cannot change our minds or correct errors, but to do that we have to take into account the harmed interests of those who acted on the basis of the previous assurances that, for instance, this was the county's long-term waste disposal plan. We thus need good reasons, especially good moral reasons, to change our minds, in such a context. We cannot justify doing what would otherwise be morally wrong without a morally good reason.

We need to note as well the loss of trust that will ensue. Depending on how compelling our case for change, others will assess one's trustworthiness in making a commitment. Having changed one's position once, when it seemed firm enough for others to rely on it and put their interests at risk, one risks, in changing one's position, a loss in trust. Others can no longer rely even on one's firm assurances of a commitment.[30]

Past decisions and present practices thus are difficult to change for moral reasons as well as for the practical reasons that habits are hard to change in any event and changes in practice are difficult to motivate. It is partly for these reasons, as I argue later, that in morality as in politics it is best to make the right decision the first time. But it is also, obviously, difficult to be sure that we are making the best decisions. Our present decisions may have a profoundly detrimental effect on our health and our environment.

We do not need to dip very deeply into the literature—no deeper than our newspapers or local television—to come on assurances that this or that particular landfill will not in any way pollute the ground water only to discover that it closed later and that the wells polluted by it closed as well.[31] The Detroit incinerator is an excellent example—touted as the latest in safe equipment, closed in less than a year for retrofitting of better scrubbers because of high levels of emission of mercury.

So the possible impact of our acts and omissions implies caution. When combined with the practices they create, more caution is warranted. Whatever decisions we make now are being made without all relevant knowledge, and thus we may well be doing something that will cause great harm. If we entrench our current acts—with heavy capital investments, for instance—we will entrench those harms. From these emerges yet another feature of the situation we are in regarding many environmental matters: We should act with great caution.

In laying out how we can be reasonable and unreasonable in gambling situations, where we lack all relevant information by the nature of the game we play, I isolated the responses on the extreme ends—maximax or maximin, unbounded optimism or deep pessimism. Rejecting both leaves us with a wide range of possibilities between the extremes, and the need to be cautious tells us that in regard to environmental matters, we should act less like high fliers and more like deep pessimists.[32]

This admonition is motivated by the great harm that can come from making mistakes in regard to environmental issues, but it is also not yet useful. For it does not tell us *how* to act with more caution rather than less. Where in that range of decision-procedures between maximax and maximin are we to settle down? Great caution tells us to settle on a decision-procedure or set of procedures that highlight the pessimistic response of maximin, and that is all. When we come in Chapter 6 to look at clear-enough cases where action is justified and where action is not, we will be in a better position to lay out the details of what, rationally and morally, we ought to do. Even then we will not end up with any decision-procedure as tightly packaged and presented as maximin or maximax: The complexities and uncertainties of the real world are not so easily pigeonholed.[33] A set of strategic principles—such as "Act with more caution rather than less!"—may be the best we can hope for, but those are enough, I think, when combined with clear-enough cases of when action is justified and when it is not.

The need to be cautious is reinforced by another feature of the situation, one common, I think, to all major public policy issues. Informally put, "Once is enough!" or "Do it right the first time!"

Major public policy issues require vast expenditures of time and energy not just because great amounts of data must be gathered and sorted and arranged but also because the criteria for decision making seem so amorphous that we feel drained by the need to get them all straight while having little idea what that means in the context. In such issues disagreements also run deep and tensions high, and that is draining. There is thus enormous pressure from elected officials, among others, to do it only once. It is dangerous territory, with political land mines; no one wants to enter; and if they must, they want in and out, only once, with a minimum of political damage.

This is especially so given the practices we have created. Major commitments, once made, have a momentum of their own, creating precedents and states of affairs difficult to change. It is thus a political truth, if we may call it that, reinforcing the need for caution about what is chosen, that what we choose will carry extra weight, and so be hard to change, because the process of deciding itself is so time-consuming and can be so traumatic on the body politic. Anyone who has sat through county commission hearings on a landfill, seen a decision reached, only to have the same issue raised again at a later date because of a change in commission membership, for instance, knows something of the frustration one feels as one must yet again enter the political thicket.

The need for caution and the desire for things to be done right the first time around both provide explanations for why decisions are difficult to change, even if mistaken, but the explanations differ. The former tells us something about what decision-*procedure* to use; the latter tells us that we need to note features of *deciding* itself, whatever the procedure we adopt. The former tells us that in deciding *what* to do, we should act cautiously: More caution, rather than less, should modify our choice so that we pick and choose among options with more pessimism than optimism about how things will fare. We should choose alternatives whose outcome leans toward providing the least bad situation, whatever happens, because once a decision is made, people will act on that decision, creating interests that would be set back if the old decision were rescinded. Our desire for doing it right the first time tells us that even if we come to see that we erred, we should have to take into account in deciding what to do not just the entrenched expectations of those affected by the prior decision but also the difficulties of again launching into a decision process to come to a new choice. We cannot just change the decision but must go through the entire process again.

The Best That We Can Do

I have been describing various features relevant to making public policy decisions about the environment. Three of those refer to us—the appropriate attitudes we should have (more caution rather than less, and do it right the first time) and our epistemic situation, or the state of our knowledge (not full now and with no likelihood of being full any time soon). The other three features describe not us, but the state of the world of environmental matters, as it were, the situation we find ourselves in—that we are in a process that will continue unabated whether we decide or refuse to decide, in which any decisions we make or fail to make have the potential for profoundly detrimental effects on our health and environment, and in which any decisions we make and practices we encourage will weigh heavily against future change.

Yet, both our epistemic state and the situation we are in are worse than so far suggested. Let us consider the situation first by sketching some of the trade-offs involved in incineration, for instance. I assume that whatever we ultimately do about waste, we shall incinerate some, but the following sort of difficulty permeates all alternatives: We are in a no-win situation.

Consider what happens in incineration. Not only will any elemental constituents such as lead that enter the incineration process also exit from that process unchanged, we will end up with new toxic compounds created by the process of incineration itself, after combustion. We can try to minimize the dispersal of these elements and compounds in the atmosphere by recovery techniques. We shall then have the toxic material in a more highly concentrated form in the ash. Or we can try to avoid the risk of concentrating the toxic materials and the expense of disposing of these now concentrated wastes by *not* recovering them. We shall then disperse them through the incineration process out the smoke stacks, far and wide. To put it perhaps much too crudely, we have a choice between concentrated globs of toxic material that we cannot incinerate and that we can landfill only at high cost and risk, on the one hand, and, on the other, a thin layer of toxic material, spread across the land wherever the prevailing winds, if any, take what goes out the chimney. The alternative to acid rain, in short, is just a less widely polluted environment, for we shall have concentrated waste that we do not handle well.

This scenario can be pursued. If we landfill the concentrated toxic remains, we can cut the costs by mixing it with non-toxic matter and so diluting it that it can be classified as non-hazardous. But although that mixing diminishes the risk of any one spot being highly contaminated, it also spreads risk over a greater area and increases the likelihood of a greater area being contaminated. A gain at one point entails a loss at another.

We are, in short, in a no-win situation regarding waste: No alternative solves all our problems, and each alternative creates new waste problems that, in turn, can be solved only by producing new problems.[34] This conclusion should not surprise us, for it is the usual situation in thorny public policy matters. There are no free lunches: The best we can do is to minimize our losses.

That last point tells us that we should not refuse to act when faced with certain losses in the hope that some alternative will eventually emerge in which we will have no losses. Because a gain at one point entails a loss at another, no matter what we choose, the best we can hope for is to minimize our losses.

But that makes it sound as though it is simply a casual restatement of the maximin rule, and it is not. The maximin rule is appropriately used only where we cannot calculate the likelihood of alternative outcomes and so presume them equally likely. But in situations where we can calculate probabilities in even a rough way, acting *only* to minimize our losses might make a loss much more likely than a great gain and so be unreasonable. Being unable to do anything but minimize our losses does not operate in an epistemic vacuum: It presumes that we can, at least roughly, calculate the probability of at least some alternative outcomes and tells us that, when we do, we shall find no alternative that does not entail losses. But we may then find that, all things considered, the best choice does not mean ignoring possible gains. According to the maximin rule, minimizing one's losses takes precedence over any gains. But to recognize that we are in a no-win situation is to recognize that no alternative is without losses, not to suggest that we should ignore potential gains in deciding what to do when we can calculate probabilities, even roughly, and so weigh gains against losses.

Indeed, if we were simply in a no-win situation and could not calculate probabilities, we might as well be high fliers if we were not overly concerned with possible harms: If there will be losses in any case, go for the maximum gain. Yet, if we can calculate probabilities, these ought to be taken into account, and if we face alternatives with potential gains, these ought to be taken into account. So even if the best we can do is to minimize our losses, we would not be using the maximin rule. For it does not tell us how to decide what *counts* as minimizing our losses— how to calculate in gains versus losses, and probability calculations, in attempting to act cautiously. For more specificity we need to wait until Chapter 6, and even then we will not get a tidy decision-procedure such as the maximin rule but some complicated cases that are clear enough to give us guidance.

We do have a further reason for attempting to minimize our losses, and that concerns the state of our knowledge. To say that we lack all relevant information is to say what now may seem obvious. "Of course," one wants to say, "we don't know *everything* that may be relevant." But that we do not know everything that may be relevant implies that we know some things, but not others, and leaves open the possibility that at least we know what we do not know and need to know. The earlier examples from gambling encourage these implications. In poker, for instance, we know what cards we hold and know what others are willing to bet. We also know what we need to know if we wished to ensure our winning, namely, what cards they hold, and we know as well that we do not know what their cards are.

That is an excellent position to be in generally, for we can direct our research at exactly what we do not know but need to know, secure in what we already know. But our situation regarding environmental issues, and waste in particular, is disanalogous in at least two ways. First, *we are not secure in what we already think we know*. One aim of incinerating trash is to destroy substances that are toxic and control those that cannot be incinerated. The operating presumption is that the process itself does not produce any new hazardous products. But if some toxic compounds form at relatively low temperatures *after* combustion, then what we thought we knew we cannot be sure about any more. It is like playing poker thinking one-eyed jacks are wild and then discovering, midway through play, that perhaps they are not. Clearly such examples are not isolated. Second, *we do not know what we need to know.* Again, PBB is a clear case. As I remarked, it is not likely ever to have occurred to anyone that PBB might enter the food chain. We did not imagine that we would need to know how to identify it when mixed with grain or when deposited in tissues of the body. We did not think we would need to know how an organism would react to it—whether it would accumulate in the body or not, whether it might form new compounds within the body, and so on. We did not think we needed to know its long-term effects on human health.

The analogy with poker helps. While playing we not only know what we do not know, but we can also rank what we need to know. What we need to know, first, is what cards the others hold, discard, and are dealt; second, what each thinks is held by us; third, what their betting strategies are in general and in this hand in particular; and so on. If we are to cheat in poker, in other words, the priorities are clear.

But it is almost correct to say that just as laissez-faire economic decision making has been driving our disposal of waste, so laissez-faire investigations have been powering our research. New technologies create new questions, accidents such as PBB getting in the food chain create investigatory problems, new compounds create a multitude of new possible combinations with old compounds and thus new possible sources of concern. Research thus tends to be more responsive than directive, and we can barely begin to rank alternatives. Besides, for anyone to try to demarcate the areas of ignorance and rank what needs to be known is to assume already some system of preferences, incineration over landfilling, for instance, so that issues involving the formation of compounds in combustion rank higher than investigation into the capacity of ground water to cleanse itself of contaminants.

To raise these questions is to enter into the sociology of scientific research—what it is that powers scientific investigations, allows some projects to be funded and others not, ranks areas of concern for graduate students, justifies some conference topics and not others. As one quickly discovers, there is no substantive agreement among those in any particular field about what we need to know. What may sound like essential agreement begins to dissipate when specific projects begin to get ranked. A listing of research priorities is thus more a mark of preference for a particular way of looking at a problem or for a particular solution than an objective listing or even a statement of agreement among investigators.

We may sum up the last two points we raised in another feature of our state of knowledge: We do not know what we need to know and may not know what we think we know. We need only think of our vast ignorance about how pollutants move in ground water, through various kinds of soils, to have a good example of something we need to know but do not know.[35] To have a good example of something we thought we knew, but did not, we need only remind ourselves, again, of the discovery of some toxic compounds formed after combustion, at relatively low temperatures. For we thought we knew that combustion at high temperatures would effectively reduce all toxic material, not produce new toxic substances. Not only do we not know everything that may be relevant, which is not a particularly surprising state of affairs, but we often do not know what we know we need to know or what it is we think we know.[36]

All this may make for a depressing state of affairs. To achieve any goal, from opening a door to becoming happy, we are clearly in the best position if we know what we wish to achieve, where we currently are in relation to that goal, and how to get from where we are to where we want to be. We are best situated, in short, when we know the *end*, the *start*, and the *means* from the one to the other. So far I have been concentrating on where we are, barely referring, without discussion, to goals of public health or of a clean environment and only mentioning in passing possible means, such as landfilling, or incineration, to achieve them. If an examination of the ends and the means uncovers as many complications as our examination of where we are has uncovered, then, one may well think, we really should despair.

Yet, we should despair only if we think that we cannot make fully rational decisions without all the relevant information or that, somehow, we cannot make reasonable decisions about how to change matters as matters change. The situation we find ourselves in regarding waste, for instance, is not unusual. It is rather the *usual* situation. We must act without even the hope of all relevant information, unsure that we have not overlooked something we ought to know but do not, and perhaps, as may be the usual case, thinking we know something that is in fact false. What we do know is that any major decision we make has the double risk that it may cause immense harm to our health and to the environment and that, once made and put into place, it will be difficult to change. Again, that is no different. Mistakes in a game, for instance, are irretrievable and sometimes fatal to the outcome.

All this implies, again, what it usually implies in such a situation. We should continue to try to obtain as much knowledge as we can and act with caution, taking especial care, as in a game, not to make any move that boxes us in but, rather, moves that open up new possibilities and take advantage of the existing flow. Yet, what can be done in any situation depends upon the players as well, and we need to examine how the decision-making process may affect decisions.

CHAPTER FOUR

Who Decides?

If we are to move in a reasonable way from where we are to where we want to be, keeping in mind the awkward epistemic position we find ourselves in regarding environmental issues and waste management in particular, we need to examine in some detail the process through which any decisions will be made—how those who decide tend to make decisions, how the decision-making process itself may affect the decision ultimately made, what sorts of issues tend to get into that process, and where in that process they make their entry. For just as our ongoing natural social artifact of waste management may get in the way of our making the best of all possible choices regarding waste, so the way we make decisions about waste may skew the choices available to us as well as the selection from those choices.

Such choices are made through our political process, and how, it may be asked, could our political process *not* skew our choices? It is a complex decision-procedure, itself a natural social artifact responding to all sorts of concerns, with all sorts of unintended effects, and it seems almost rhetorical to ask what the likelihood is of its producing the best choice among the array of possibilities for any particular environmental problem. With a design that allows it to respond to a wide variety of problems, affected by an enormous number of different causal factors, it is as unlikely as flipping a coin to produce the best choice among the different options available for handling waste, for instance. Indeed, the analogy is inappropriate because flipping a coin might produce the right answer, whereas, I shall argue, the decision-making process will tend to skew the choice.

The lesson is not that we need a czar of waste, for instance, but that to the extent that we do not back off from the decision-making process so we can understand it and its implications we shall be captive of the way it "naturally" makes us think in skewing the array of possible options and possible solutions. We must obtain a realistic sense of what is possible, recognizing that no political process of decision making would have produced an Eden, and strive for as much leverage on the system as we can, encouraging those natural tendencies that are helpful and altering, where we can, those tendencies that run counter to making the best choice.

The issue is thus not whether the process will alter the array of choices but how it will do so and, therefore, what are the most significant ways in which it will do so given our concerns. I will concentrate on four interconnected features of the political decision-making process that, I argue, make a difference to our concerns:

1. How it tends to transform all concerns into interests so that decision-making becomes a matter of weighing interests against one another
2. How long-term problems that only gradually become prominent tend to be ignored until the damage they have caused cannot readily be reversed
3. How the solutions tend to be those in which the bad effects are far removed and the good effects are more immediate, that is, before the next election
4. How politicians tend to inoculate themselves against the effects of their own decisions, especially where some knowledge of science is necessary.

Factors other than these four enter into the complex process of political decision making, and in fastening on these causal factors, I am not claiming they are all that count.[1] As we will see, other factors encourage—and yet other factors discourage—just the sorts of skewing I think we need to guard ourselves against. I do not aim to provide a Rosetta stone of politics or to lay out in its entirety the complex natural social artifact that is our political system but to emphasize features of political decision-making I think tend to be ignored and yet are of crucial importance if we are to shift the way in which we think about environmental issues and waste management in particular. I think it helps to look at decision making from the point of view of those making the decisions,[2] and so I begin with some remarks about what it means to talk about what a politician, as a politician, tends to do.

Professional Positions

Those in different professional positions in society—lawyers, nurses, physicians, professors—have their own ways of conceiving of and so approaching problems, their own particular attitudes, and their own modes of behavior built into the positions they occupy. Professionals come to occupy their positions by training, professionally now in most cases, or by a kind of apprenticeship, as can occur in politics. A person fully occupying a professional position acts as a professional when the act is marked by the principles that define and sustain that position.

A brief illustration may help, for at least three points are being made: that such professional positions are defined by sets of principles, that the sorts of principles differ in different professional positions, and that actions (or omissions) from within that professional position differ accordingly. That one ought to care for the ultimate well-being of one's customers is no part of the defining criteria of business, but it is one of the principles of nursing. If a business were to act on that principle, then, although the novelty of the enterprise might attract customers, it is unlikely in the long run that the business would succeed competitively. But a nurse who did not act for the long-term health interests of patients would not last long, and neither would the patients.

All this can go very deeply into ways of relating to others. A friend tells a story about sitting in his office in the medical building of the university at which he teaches, hearing tittering from down the hall. He ignored it for awhile, but it continued, distracting him, came from more than one source, and seemed oddly out of place. So he investigated only to discover medical students trying to learn how to examine each other—without tittering. The last thing we want when a physician asks us to disrobe is that he or she titter, blush, or laugh outright. Yet, embarrassment is a spontaneous response to nudity, in our society at least, and so those in professional positions who must deal with nude persons need to be trained out of that response or trained not to show it. They need to come to see the body as a mechanism, more like a mechanic looking at a bicycle than like a voyeur.

To describe how people are professionalized into a certain attitude and into certain modes of behavior is not to criticize that professional response. If someone becomes a professional pickpocket or a professional misanthrope, criticism is appropriate. But merely acting as a professional—a physician, a politician, or a nurse—is not in itself cause for criticism.[3] Difficulties can certainly arise, however, when a professional acts professionally outside the appropriate sphere. A lawyer trained to press clients to ferret out damaging information to the client's case will find that this nonempathetic attitude will not serve well in family matters. There we want empathy. So a professional needs to be aware of how becoming a professional may have shaped an attitude toward the world and be careful not to extend it beyond its proper bounds. A nurse cannot care for everybody, but when acting as a nurse, he or she is not subject to criticism for doing so.

Thus, when I say that a person acts as a politician, a public administrator, or a political appointee I do not mean these phrases to have the pejorative tone they may carry for some. Politicians are professionally politicized, public administrators are professionally bureaucratic, and so on. I mean only to describe some of what a politician tends, as a politician, to do, the nature of the decision-procedure a politician tends to adopt, and so the kind of decision a politician tends to make. But these descriptions are meant to carry a weight that sociological generalizations do not have, for they lay out what may be called the depth grammar of political decision making—how politicians *must* decide to be politicians, how the nature of the decision-procedure *must* work in a political atmosphere. Like any description trying to provide the deep structure that underlies a practice, my descriptions are normative.

I thus risk misunderstanding in two ways at least. I may, first, be thought to be confusing how the politician *tends* to decide *as a politician* with how a politician *ought* to decide, given what we may call the role-morality of politicians. I intend to cut through that distinction: The role-morality of a politician just is what a politician must do, as a politician, and so tends to do—and only *tends* to do because, luckily perhaps, few politicians or professionals of any sort always act purely *as* politicians or professionals.

The underlying cast of mind here is David Hume's. In his essay "That Politics may be reduced to a Science," he argues that there are laws in politics as certain as any in "the mathematical sciences."[4] He then proceeds in his political and economic essays to examine some of those laws and their implications. In "Of Public Credit," for instance, he notes as one of the laws of modern politics that the "practice . . . of contracting debts will almost infallibly be abused, in every government."[5] The Romans, he says, carefully paid back quickly whatever they had borrowed, but "our modern expedient . . . is to mortgage the public revenues, and to trust that posterity will pay off the incumbrances contracted by their ancestors."[6] He finds an explanation in the political character: "Why," he asks, "would a minister [in Her Majesty's government in Great Britain] persevere in a measure so disagreeable to all parties [as to tax them] in times of peace and security, when alone it is possible to pay debt? 'Tis not likely, [he adds] we shall ever find any minister so bad a politician. With regard to these narrow destructive maxims of politics, all ministers are expert enough."[7] His point is that in times of war and turmoil, politicians can justify increasing the public debt because of the necessity of raising money to meet the current crisis, but when the time of crisis is past and the debt remains, no good politician will act to put popularity in jeopardy by raising taxes.

I give a similar sort of analysis in Chapter 1 when I argue that the most advantageous choice for a politician facing an election is that in which the benefits are weighed toward the immediate future, before the next election, and the burdens come far removed from the time the politician is likely to be in office.[8] My example and Hume's are meant to be instances of the role-morality of politicians, instances of what politicians must do as politicians and thus of what politicians tend to do. That "must do" is meant to mark that these claims are not just descriptions of what politicians generally do but are constitutive of the practice of being a politician.

I thus also risk, secondly, the reader's patience. Descriptions that purport to lay out the deep structure of any practice, from how scientific investigation must go to how a novel must be written to *be* a novel,[9] sound—and often are—pretentious, as though the author had access to a level of reality different from that we all move in, as though the descriptions were meant to be immune from criticism. How one does criticize such descriptions is a difficult philosophical matter, but I do not intend my descriptions to be immune from criticism. I offer them with more than moderate Humean skepticism, convinced that our forms of political relation, among our various forms of relating to one another, are conventional, however stable, and aware of how readily we humans tend to think ourselves immune from the sort of epistemic shortfalls I have examined.

What I have in mind is what cultural anthropologists strive to achieve, ever aware of the real possibility, and even likelihood, of failure because "it is hard to be conscious of the eyes through which one looks."[10] Just as the aim of a cultural anthropologist is "to describe deeply entrenched attitudes of thoughts and behavior" of a people, so my aim is to lay bare the deeply entrenched principles of a practice.[11] What I should hope for is what a cultural anthropologist hopes for, that anyone whose practice is described would find the principles so obvious that they would hardly think them worth remarking on. How could any politician not recognize the truth of Hume's remarks about the political difficulties of raising taxes to pay off past debts when everyone is otherwise satisfied?[12]

I also have modest ends in mind. First, I am concerned with those aspects of decision making that make a difference to how decisions about the environment and about waste are made. Some general principles I hope to articulate are descriptive of more than simply politicians as I am concerned about the decision processes used by bureaucrats and political appointees as well, but I am not concerned to lay out the general principles of political decision making.

Second, in any democratic society at least, one is luckily plagued by political amateurs. That is a burden to a theoretician who would like to delineate the pure professional politician but a benefit to all of us for some of those reasons that make democracy a benefit, fresh ideas, idealism, and so on. So to say of someone that he or she is a politician is not to say, in our society, that he or she is a professional. It may only mean that they hold political office. The principles I articulate about political decision making must be tempered accordingly when applied to actual decision makers.

Third, I am not concerned with every trait or attitude of politicians or in laying out fully "the political character" but only with some of those features of political decision-making that affect political choices. If we are aware of how a professional position may affect the way we look at a problem as well as how, because of our training or apprenticeship, we will tend to solve it, we will be better equipped to determine if that choice is the correct choice or is only "naturally" what someone in that position would choose. A politician may choose an alternative where the benefits occur immediately and the costs are borne by future generations, but that choice, a natural one for someone acting in a political capacity, concerned about the next election, obviously may not be the best choice, all things considered.

I do not imply that "a natural political choice" is incorrect per se, only that it needs to be assessed by broader standards—as should any professional decision. A dentist who makes a professional judgment that one ought to have one's wisdom teeth pulled cannot expect a patient who cannot afford it to have it done merely because of the professional judgment. Other values can and should enter into such a calculation, and although we have delegated public policy decisions to political bodies in our society, that does not make them immune to criticism in accordance with values squeezed out by that political process.

Political Decision Making

The electoral process requires a politician, to be electable, that is, to appeal to broad interests of a constituency.[13] Being re-elected is not the only interest a politician has. Any president in a second term is a counterexample to that thesis, but it is *an* interest and one that, with a variety of other causal factors in our society, has made the currency of political decision-making interests, not rights.[14] Claims to rights that run or seem to run counter to the public good tend to be ignored or perverted, and although rights can and are defended in the abstract, a politician can defend individual rights that clash with majority interests only with care to avoid unnecessarily antagonizing the interests of a constituency. We call those groups which provide the finances necessary to run a re-election campaign "special interest groups," and the name is not an idle one. It emphasizes that interests are the currency of politics—in more ways than one.

This concentration on interests, especially those particularly relevant to a politician's chances for re-election, has several implications that are particularly harmful to the chances of environmental issues being considered and, once considered, settled in a way most appropriate for them. For many environmental issues only gradually come to be perceived by the public as problems—as toxic wastes slowly accumulate to a critical mass, or as the trash we each produce mounts each year and our dump sites begin to fill to the point of overflowing, and so on. It is only when a problem has reached such a critical mass that enough people are affected by it to bend the appropriate political ears that action is considered, but by then much damage, some no doubt irreversible, may already have been done. In addition, politicians also tend to immunize themselves from the consequences of decisions by foisting them on others when they can. One effect of the issues about waste management requiring scientific knowledge is that scientists are asked to make recommendations, and we then enter, I shall argue, a black hole in which few clear recommendations can emerge. The selection from those that do is skewed by politicians' tending to choose that alternative in which the goods are front-loaded, as it were, so that the electorate can see that they have gotten something beneficial for their tax dollars. The preferred alternatives are those in which the good effects occur before the next election and the bad effects a long time after. The selection of alternatives for waste management is thus skewed by the process we must use to select them. That process provokes less than the best decisions.

I do not think I say anything surprising in noting these various aspects of political decision making, but I am not providing a comprehensive theory in terms of which to comprehend and rank them. My interest is only in laying them out in enough detail for our purposes. Let me begin with interest-resolution.[15]

Interest-Resolution

In his "Idea of a Perfect Commonwealth," David Hume noted that one of the advantages of having a large republic is that factions—what we call special interest groups—would be so spread out that they would be unable to dominate the democratic assemblies as they might in a town meeting or an Athenian democracy.[16] Madison advanced the same argument in *Federalist Paper* Number 10,[17] and we have come to see that Hume and Madison were more hopeful than accurate, especially given a political climate where two major parties are relatively equally balanced so that a small group can have disproportionate influence and given modern methods of mass communication that allow small factions readier access to more citizens. A small interest group can dominate a political forum as well as a political agenda. Although not a majority, such a group may control enough of a percentage of the electorate to tip an election, and if the group votes as though that issue were the only issue, politicians, quite rightly, must take care. Special interests become especially important.

In short, the political process accentuates interests. One consequence is that, in concentrating on competing interests, it tends to ignore relevant rights. It is helpful to conceive of rights here as interests that, for a variety of reasons, we deem so important that we elevate them out of the political calculus of interests into a realm of their own. There they compete with each other, but not with interests: Those they trump. Rights in conflict with interests ideally always win. If there is a right to desegregated schools, for instance, an appeal to an interest in having neighborhood schools is irrelevant. If we have only an interest in neighborhood schools, that appeal cannot even begin to weigh against a right to desegregated ones.[18]

The right to free speech is a right guaranteed in the First Amendment. That means that even if it is in the interest of a majority that some person's or group's right to free speech should be denied, that right ought not to be denied. A right implies that a democratic majority is not to have its way regarding the issue at question. A right is meant to immunize certain interests not only from factions but also from majorities.

It is just the point and the problem of such rights that they can clash with the majority will and that their consideration tends to be squeezed out by the political process. When they are asserted abstractly, they are in the interests of no special interest group in a political process that tends to accentuate special interests. When they are asserted as the rights of a particular group, so that they are not abstract, but real, the political process converts them into real *interests*. So if it is the particular function of political bodies such as county commissions, state legislatures, city councils, and so on to determine public policy, and to do so by representing the public and thus, presumptively, the public interest, consideration of how that representation may affect rights tends to be put to one side.

It is easy enough to think of examples where this has both occurred and has not. The rhetoric of rights can get in the way here. It seems far too often the case, or more cautiously but depressingly we *suspect* it is far too often the case, that politicians have falsely claimed to take the high road—arguing for a position, taken for reasons of interest, by claiming it is made out of concern for right. That we suspect that politicians try to mislead us in this way is symptomatic of the claim I am making that a tendency exists in the political process to squeeze out consideration of rights. To the extent that political bodies see themselves and are perceived by their constituents as arbiters of interests, to that extent rights will be ignored.

I do not want to claim here that there is a right to have an environment clear of hazardous waste and unsullied by it or to speculate on how we might actually try to insinuate such a right into the decision-making process without resorting to the judicial system but to consider how such a claim would fare in the political process. If it were accepted, its impact, I think, would be clear enough. It would require that the water in our homes not be contaminated by lead and thus would require, for example, that cities such as Chicago spend whatever sums are necessary to replace the lead pipes that were required by law until 1986. It would preclude, for instance, acid rain—if, as seems more than simply likely, it is acid rain that is responsible for the loss of so much of our natural environment in the northeastern United States and in Canada. If as many lakes and ponds and trees and shrubs are dead as a result of acid rain as is claimed, then a right to have an unsullied environment would presumably outweigh whatever interests there are, usually economic, in maintaining the conditions for such damage.

Yet, this form of argument and this kind of consideration tend to be as irrelevant in the political forum as an appeal to an interest in local schools would be in regard to a right to desegregated schools. The point is not that there is no one to represent that right but that an appeal to that right is the wrong sort of move to make in the practice of politics.

Better, an appeal to a right *becomes* an appeal to an interest. Introducing a right into the political process is not like trying to jump a rook while playing chess and having someone point out that jumping pieces is what one does in checkers, not chess. The jumping there is simply not a move in the game. But if the right to an unsullied environment becomes an object of concern to any interest group, so that they can present the appeal to their representatives, then that appeal will, by the nature of the process, be transformed into an appeal to interest. Either rights are squeezed out of the political process as irrelevant or they become relevant by being transformed into interests.

When such a transformation occurs because of the structure of the political process, it is difficult to see that anything has gone wrong. Courts do exactly the opposite, but with the same result that it is difficult to see that anything has gone wrong. We may go to court to resolve competing interests, but the very nature of the judicial process converts interests into rights claims. Anyone who has been through a divorce is familiar with this process. Two parties, with sometimes competing and sometimes compatible interests, are forced, by the nature of the procedure, to make claims against each other to be adjudicated by the court. Any attempt to accommodate their interests one to the other and work out some compromise is retarded by the judicial process itself, which demands competing claims to be adjudicated. We will hardly notice the change that the structure of the procedure forces on what were, originally, competing interests. They have been silently transmogrified into rights.[19]

In a similar way, the political process transforms claims of right into considerations of interest, and, again, because it is the structure of the process that produces the result and we operate within the normative constraints of that structure, the transformation is difficult to notice. Rights are essentially changed so that they can no longer function as trumping claims.

On the one hand, the judicial process transforms interests into rights, and on the other, the political process transforms rights into interests. There are three aspects to the transformation of interests into rights by the judicial process. First, what would have been considered interests, for example, are *represented* as rights that would, normally, trump the interests of the other party. Each party to the dispute is jockeying to have its concerns expressed in a way that will allow them to trump the interests of the other party. Second, the judicial process itself transforms the party's interests into rights. It is not just that courts will not *recognize* anything but rights and so will not consider anything but rights, although that tends to be true and drives the parties to put forward for resolution rights a court clearly cannot refuse to consider. Rather, what would have been considered interests in the political arena are transmogrified into rights by the judicial process, with the additional weight rights carry. Third, the parties emphasize those aspects of the dispute that can be adjudicated and carry the most weight. Interests tend to fall out, and rights that might otherwise not have been of major concern come to the fore.

We need only examine various conflicts over waste incinerators or dump sites to see this process occur.[20] Once the parties to the conflict consider entering the judicial arena, they both (come to) have rights that courts must decide between and (come to) put forward rights that they think have been denied, emphasizing them over interests they may have that they must be convinced, I would argue, will not be treated as rights and so properly weighed by the courts.

It is arguable that the structure of arguments about rights aids and abets this transformation. After all, rights, like interests, can be matters of degree so that some are treated as absolute and others as prima facie, subject to being overridden if other relevant rights are more weighty in the context. In short, rights compete with one another like interests so that the most weighty right, like the most weighty interest, is supposed to win any competition, and so on. It is arguable that rights and interests are easily transmogrified, the one into the other and vice versa, because their essential structural characteristics, as marked by the way they function in political and moral arguments, are so similar. In any event, just as the structure of the judicial process tends to make an accommodation of interests impossible, so the political process tends to make it difficult to appeal to rights.

One consequence is that political discussion about environmental matters tends to be intellectually impoverished. No appeal to rights remains untarnished by the political process, and to the extent such an appeal would make a difference, the transformation impoverishes the discussion. An impoverished discussion is as likely to lead to a poor conclusion as an impoverished diet to malnutrition.

A second consequence is that the form of political discussion will of necessity be a weighing of competing interests. No argument of right can be made that would trump any interests. Representatives would be ill-advised to trade their constituents' rights, garnering a favorable vote for one right by giving up a vote on another, but that sort of trading occurs all the time in democratically elected bodies, and it can occur—logically is possible, I want to say, without the moral inappropriateness that would occur if rights were being traded—simply because the very form of political decision making means that politicians weigh interests and that interests are not of so much concern to those being represented that they cannot be traded one for another.

The third consequence is that any decision coming through the political process will be skewed to just the extent that it ought to reflect considerations of right and does not. If, for instance, the crucial question about our management of waste is whether we all have a right to an environment clean of waste and its effects, then that question will not figure as the crucial question by the nature of the process. It can at best be one consideration of interest among others.

For instance, most citizens seem not to want to have hazardous waste treatment facilities, dumps, incinerators, and so on in their backyards,[21] and so, in the face of such opposition, much research has been done on what would have to be done to facilitate such siting. An incinerator, for instance, has to go *somewhere*, and because those in the vicinity of its preferred location are voters so that any politician would prefer their being cultivated and convinced to accept such siting, much research has been devoted to what form of consideration would produce acceptance.

The form of the enquiry is straightforward. Those who oppose siting an incinerator near their homes "are being asked to bear high personal costs (in the form of risks) while the benefits of the facility accrue to a larger outside population."[22] So the question is what consideration would be sufficient to make them willing bear those costs.

One suggestion is that they be given economic benefits sufficient to overcome their aversion to accepting such a risk. If they accrue benefits that the "larger outside population" does not receive, then "the imbalance would be redressed," it is claimed, "and public opposition would abate."[23]

In fact, such incentives have not generally been effective, and in response to their failure, it has been suggested that what is needed is "risk substitution." The aim is to look "for ways that treatment facilities can be sited without significantly worsening individuals' *perceptions* of the risks they face in their everyday lives—and perhaps looking for ways to work toward actually diminishing peoples' real and perceived risks." So, for instance, "a hazardous waste incinerator might be located at the site of a Superfund hazardous waste site with the expressed purpose of cleaning up the site. Or a chemical treatment facility might be located at the site of a troubled chemical manufacturing plant." The claim is that because the citizens in those areas already are at risk and, presumably, perceive that they are at risk, the aim should be to change their risks, actually working to diminish them when possible in what is called "maximal risk substitution" and adding "no danger to existing risks" in "minimal risk substitution."[24]

This process is meant to be a gain on the technique of economic compensation "where people perceive that they are being 'bought off'."[25] But it has its own problems. On the one hand, it is not clear why anyone who happens to live in an area abandoned, say, by a chemical manufacturing plant would be willing to accept as a solution a chemical treatment facility that would remain long after the residue of the plant was cleaned up. It seems far more reasonable to opt for a solution in which the plant residue is cleaned up and the land left pristine for future, non-risky, development. Indeed, it is only if there is the additional premise that nothing will be done to the residue of the plant *unless* those near it accept a chemical treatment facility that the concept of risk substitution seems to have any leverage. Why should voters be willing to substitute one risk for another when, they may reasonably think, they should not be at risk at all? Whereas providing economic incentives looks like bribery, substituting one risk for another looks like bullying. So people may not think they are being bought off but threatened.

On the other hand, the discussions of what form of consideration would convince people to accept an incinerator nestles within a conception of the nature of the problem that already concedes what some may consider the most important issue, namely, whether those who live near a toxic dump site, for instance, have a *right* to have that cleaned up without having to pay for it by having a new source of risk enter their lives. It is assumed that what are at issue are competing interests—the interests of citizens not to have a new source of risk in the form of a hazardous waste treatment site, the interests of other citizens not to have to take on an additional risk for the benefit of the rest, and so on. It is only within the confines of that understanding of the problem that trying to change people's perceptions of risks as well as their risks makes sense. The underlying premise is that any solution will require their giving up some interests to gain others—their giving up hope of not facing a certain new risk "in their everyday lives" for getting rid of a risk for which they may bear no responsibility. It is presumed, without argument, that is, that the question of whether they have a right not to have such a risk is moot.

It has been squeezed out of consideration by the process of decision making or transformed by it into just another set of interests—to be bought off in some way or bullied out of, by compensating interests. It is another question, of course, whether there *is* a right not to have to suffer the risks associated, say, with a chemical dump site left by a factory that has closed its doors, but whatever the answer to that question, we should not think that the relative absence of rights from political discussion about waste is a function of their lack of relevance or their lack of power. Rather, it is a function of the structure of decision making in our political process. What is at issue is not how to alter people's perceptions of risk but the conception we have of how to make decisions regarding such risks.

Long-Term Problems

The political process will skew the creation of Eden in another way, for it also tends to force us to focus on problems that have become widespread, and serious enough to affect adversely enough interests of enough constituents that they can no longer be ignored. As I have argued, when a politician must act, he or she is likely to decide in such a way as to minimize the political damage. But often the most effective way to minimize such damage is not to consider acting. A politician's best solution is the conservative one of avoiding problems when possible. If the aim is to ensure re-election, "it makes sense . . . to follow conservative strategies."[26] Taking a position on something, especially on anything as controversial as the siting of an incinerator, is guaranteed to antagonize at least some of one's constituents. But a politician can readily get away with not acting in regard to those long-term problems that only gradually become prominent. Those sorts of problems thus tend to be ignored as long as possible, often until the damage they have caused cannot be readily reversed.

Important environmental issues fall into this class of issues where the effects are cumulative and so gradual that they have a low political profile until after they have reached a tip point where we may not easily be able to stop the slide into the worse effects. In a city like Los Angeles, in which over the years more and more cars were driven, the resulting pollution crept up, and drawing attention to the problem early on was not easy.[27] It was too far removed in the future to be considered a problem, especially when there were other more immediate and pressing problems.[28]

A good example is soil erosion. It has been estimated that "25 billion more tons of topsoil are lost to erosion each year than are formed."[29] It is difficult to do more than estimate, but no one doubts that topsoil is being eroded and that we would be better off were we able to retain more of it. It takes "200 to 1,000 years to replace one inch of topsoil," under the best of conditions, and so only an intensive concentrated long-term effort can repair any loss. Unfortunately, we are even less likely to be able to sustain such an effort through the generations it would require than we are to prevent the erosion to begin with. The difficulty is that, relatively speaking, the erosion occurs so slowly and so imperceptibly, that it is barely noticed until the losses are very high—in the distant future. When one begins, as in Iowa, with feet of topsoil, losses of even several inches a year are not noticeable, and, worse, it is in no one's immediate self-interest to act to prevent the erosion. Major and perhaps irreversible harm is so far in the future that no immediate incentive exists for preventing it. Farmers "in a poor economy . . . [who] tend to live hand to mouth . . . see little incentive to protect the land from damage that might not become apparent for another decade."[30]

What is needed is some solution to the long-term problem that also offers short-term benefits. For instance, vetiver is a grass with roots 6 to 10 feet deep that will readily hold soil in place "while the plant's tall blades of grass form a barricade against runoff above ground." It can be "planted close together to form a wind barrier," and it has the immediate benefit of keeping "large amounts of water on crops longer . . . [so they] grow faster and denser." Any farmer would be well advised to plant something like vetiver because it will help produce bigger and better crops.[31] Even someone consumed with immediate self-interest will do something, for immediate self-interest, which happens to have long-term beneficial effects.

The cumulative long-term effects of years of pollutants coming out of smokestacks present a problem whose effects could have been prevented by timely action but were so far in the future that a political process that emphasizes what matters for the next election would not notice it. Even when a risk is claimed to exist, doubt can readily exist about whether the risk is real or whether, if real, it is great enough to cause concern. The epistemological card can always be played, especially given the slow and cumulative nature of the harm that is supposedly occurring.

The political process is ill-designed to handle such a problem because it will tend not to take cognizance of it and because, if it does, it will tend, as I shall go on to argue, to opt for those solutions whose good effects are more immediate even though the long-term effects may be harmful. One way to finesse such problems is to find long-term solutions with short-term incentives—the political equivalent of vetiver.[32]

Front-Loading

It is commonplace in moral matters to distinguish what is moral from what is prudent and what is prudent in the long term from what is prudent in the short term. The distinctions are made in part because of the very strong tendency we have of weighing all issues as though our short-term interests were all that mattered. When contemplating the pleasure of an ice cream, or of a strong drink, for example, we often find it difficult to consider that the immediate pleasure may be followed by other consequences we may not wish. It seems even more difficult to separate ourselves from our own interests and consider matters from a moral point of view, from how it would affect others as well as ourselves, without weighing our own point of view more heavily than any other's.

The aim of such distinctions between what is moral and what is prudent in the long and short run is to come to grips, conceptually at least, with this very human tendency to choose those alternatives whose benefits are more immediate and self-interested relatively independently of whether the overall losses are worth the immediate gains. Because the political decision procedure we have is heavily weighed toward resolving competing interests, it is biased against any choice that expresses the pure self-interest of any participant. Coming to a decision where there are competing interests means that the participants must at least consider interests opposing their own.[33]

It is no doubt too much to hope for that any human enterprise concerned with governing would institutionalize the moral point of view, requiring it of its practitioners, even training them into that attitude.[34] However that may be, one human enterprise does seem to institutionalize a concern for what I call front-loading benefits, and that is the political enterprise as embodied in city councils, legislatures, and so on. For in the competition of interests, some interests dominate over others: Those that are of immediate benefit weigh more heavily in the calculus.

This claim may seem to those within these institutions offensive and even insulting. But to claim that such political bodies institutionalize front-loading is not to accuse them of acting immorally or to say that such political bodies cannot act morally, but that the nature of the institution, and the general tendency of those in it, is to prefer solutions in which the benefits are front-loaded—up front, relatively independently of the long-term effects.

The reason is that the next election is nearer rather than later. We may cite the electorate's tendency to forget after a brief time, their tendency, in a democratic society in which news seems dominated by transient crises, to forget after a few crises who was involved in previous ones; the electorate's concern with what they see as having been actually done by the time of the election and their distrust with what they are told will occur in the long term; the electorate's awareness that one political season's long-term plan may be another political season's boondoggle; or the electorate's realization that because most solutions take a great deal of time to come to fruition, they are not the primary beneficiaries. If they pay a heavy price up front, they risk having only those who come after them benefitting.[35] These are quite legitimate reasons for an electorate's preferring solutions in which the benefits are more immediate. Because participants in the decision-making procedure are concerned with re-election, the electorate's preferences are a good reason for their concern with front-loading.

Yet, whatever the causes, the tendency to appeal to those alternatives in which the benefits are front-loaded is a real feature of the democratic process. It reinforces what may be a natural human tendency to take the goodies up front. Any action or omission has good and bad effects, and an action or omission itself may be good or bad or a mixture of the two. If I get a vaccination, the needle going in may be painful, but the effect is usually to inoculate me, with no side effects, for a net benefit. But where the act and effects are mixed and where the act may cause pleasure and the effects may be painful, we are often inclined to act without much regard for the subsequent pain. It is difficult, otherwise, to account for the tendency of some to drink too much.[36] If it worked the other way around, if, after drinking too much, people first got enormous headaches, with their lips and throats dry, and were generally out of sorts and ill at ease with themselves for a day or so, and then got a pleasant buzz that lasted a few hours, I suspect few would drink so much.

If we could line up the alternatives for a particular problem, lay out them and their consequences, marking each with its requisite amount of good and bad, the ways in which the good and bad features might cluster would presumably have no essential connection to their overall benefits. One generally bad choice might have all its good features up front, perhaps as with drinking. Another generally good choice might have all its bad effects up front and all its good features far down the road. Nothing about the nature of alternatives for *any* sort of problem, I am suggesting, guarantees any particular configuration of good and bad effects. But, in fact, solutions to environmental problems often require immense capital expenditures now with the benefits coming only much later. So, if there is any bias in the alternative solutions themselves for environmental problems, it is a bias toward alternatives where the good effects are rather farther removed from the next election than the bad effects.

But, quite independently of that, the actual choice made by a political body, I am suggesting, will tend to be that alternative in which the good effects are more immediate, independently of the overall merits of the proposal. They may be nothing more than fewer taxes being levied right now, but, whatever they are, one consequence of entrusting decisions to the democratic process is that the calculation of interests tends to be short-circuited: Those interests which are immediate and beneficial weigh more heavily than any others. Among all the possible solutions to a problem, the process tends to highlight those in which the good effects are front-loaded.

We need only recall how difficult it is for any political body to raise taxes or slash benefits to cut a deficit in a budget. As Hume remarked, it is "not likely we shall ever find any minister so bad a politician" as to raise taxes "in times of peace and security, when alone it is possible to pay debt."[37] But it is also easy to find examples if we concentrate on environmental issues. For instance, the history of how we have dealt with oil tankers is a classic example of how we fail even to consider seriously all the alternatives because those alternatives which do not have benefits up front tend not to be considered.

There is no doubt that oil leaks are a major source of pollution. The leaks come primarily from tankers getting staked on sharp rocks and from rusty ballast tanks that corrode and cause leaks. These problems could be either solved or significantly reduced in a variety of different ways. Double hulls would help. Keeping the ballast tanks full at all times and so loading the oil that "the pressure on the inside of the hull equal[s] the pressure from seawater on the outside" would not only prevent the flow of oil out of a ship with a punctured hull but also the frequent alternation of "salt water and salty, humid air" in the ballast tanks, which causes rust. Putting "emergency shutoff valves on the air vents" would also equalize the oil pressure and water pressure so that "after a little oil leaked the vacuum pressure would hold the rest in." And a new design recommended by Mitsubishi would have ballast tanks down the sides of tankers, providing "a kind of bumper," and, more important, "stacked" oil tanks in the center so that oil pressure could be kept low, achieving the effect of equalizing external and internal pressures.[38] Yet each of these alternatives requires expensive up-front costs—brand-new ships in regard to Mitsubishi's recommendation and in regard to double-hulled ships, significantly less oil being carried each trip, or expensive and complicated emergency locks for the other.

It takes no genius of political decision making to realize that replacing the existing tankers of the world now with new tankers is a non-starter. It is not just that each country's tankers would be at a competitive disadvantage so that what is needed is international co-operation. That could be achieved with the will. Rather, the difficulty is that the solution is expensive, with enormous up-front costs and few up-front benefits, if any, to outweigh those costs. We not only would lose all the existing tankers with the huge investments made in building them but also would have enormous new expenditures for the new tankers, all of which are more expensive than current models. That we shall therefore have further oil spills, some no doubt of great magnitude, does not change the fact that the *best* we can hope for is that as existing tankers are replaced, and as new tankers are built, they are built to minimize future leaks.[39] That this is *obvious* is a mark of just how pervasive is the mode of decision making in which even the listing of what we would consider realistic alternatives is determined by a process that weighs more heavily the good effects that are front-loaded. Where there are no benefits up front to highlight, the alternative disappears *as* an alternative.

But that mode of decision making has another damaging result: It tends to produce half-way measures. Just as there now seems no sense, unfortunately, in designing and constructing buildings to last much beyond the years they accrue tax benefits to those who invest in them, so there seems little sense, to someone facing an upcoming election, to ask, for example, for great outlays of tax money to provide a solution that will survive the immediate tax concerns, let alone the tax lives of those being represented.

It is thus not thought at all unusual to plan for a low-level radioactive dump, where the half-life of the material is in the many thousands of years, and to point out, in response to a constituency concerned about environmental issues, that the dump is good for 500 years. Shrink the years, and the same sort of thinking remains about other waste material. So it is not unusual to propose a landfill that would be filled in ten to fifteen years and whose capacity to retain the wastes and protect the ground water has little greater life span. That we are now coming up against the consequences of such short-term solutions made fifteen or twenty years ago should serve as a reminder, not that we ourselves or our immediate civic ancestors were stupid or ignorant or even short-sighted, but that the way the system operates may be stacked in favor of such short-term solutions—that the system makes shortsightedness a benefit, if not a virtue.

There are not just political reasons for choosing alternatives in which the benefits are more immediate, even if the alternative provides a less than optimal solution in the long term. In making a decision, we can rarely if ever lay out ahead of time all the alternatives, with all their effects, good and bad. Even in such artificially restricted environments as chess matches, master players are surprised again and again by some unforeseen effect of a move. In the real unrestricted world, especially that of public policy, surprise ought to be presumed the rule rather than the exception. It is commonplace for a decision made with intelligence and apparent foresight, with as much information and care as were available, to have totally unforeseen effects. We dig wells in the Sahara with the best of intentions and discover that the balance of cattle to grazing land is thus upset, leading to the loss of covering vegetation and thus of the very capacity to sustain life. Perhaps we can say in retrospect that we ought to have foreseen such an effect. But we did not, and it is our usual state of affairs that far from knowing all the alternatives and their consequences, we know only some, or at least do not know if we have them all, find ourselves ignorant of much that we would need to make a godlike decision, and are thus unable to choose well among the alternatives presented in such truncated fashion.

In such a situation of ignorance and uncertainty, it does not seem unreasonable to choose that alternative in which the good effects are immediately forthcoming. The farther removed from one's actions an effect is, the more likely it is that something—a change of political parties, the collapse of the Soviet Union—may intervene to prevent it; the farther removed an effect is, the more likely it is that although it looks good now it may be bad then; and so on. "A bird in the hand is worth two in the bush" has deep political significance. So, far from castigating those politicians who make such decisions, we may want to applaud them.

Such decisions are also encouraged by our economic system. From the producer's point of view, the best product is something that everyone needs and that does not last long. The shorter-lived, and the more necessary a product, the more profitable the company that produces it. So a solution to a problem with long-term benefits—like a bar of soap that lasts a year, say—is not preferred. In short, and more generally, "the market price system is unlikely to favor far-sighted . . . investment decisions."[40]

It is true, however, that regarding some matters such as waste and pollution solutions with the main benefits farther away rather than sooner are more likely to be better than solutions in which the benefits are front-loaded. It is also true that the process of decision making tends to squeeze out such contenders as much as it transforms appeals to rights into appeals to interest. To the extent that such solutions are less likely to be chosen or even considered, the intellectual discussion tends to be harmed accordingly. Arguments about policy that depend on the long-term interests of those being represented or of society as a whole will be given short-shrift because they will be seen not to be serious contenders for political action if they do not also have short-term benefits.

The practical effect is that real political pressure exists in favor of short-term solutions—either those in which the benefits are up front, relatively independently of the long-term effects, or those which are inadequate in the long run but represent politically acceptable short-term costs, such as low taxes. How that bias built into the system can be systematically overcome is a nice question and one I shall not speculate on.[41] What all the causal factors relevant both to produce and to overcome this built-in bias are is an intriguing question. However, my long-range concern is the stance toward decision making regarding environmental issues that encourages, and is in turn encouraged by, just this bias. As we saw, among the range of options the Michigan Department of Health considered when deciding what to do about PBB, the one that clearly caused harm up front was the least favored, quite independently of such other considerations as the long-term effects of banning or not banning the foodstuffs. Changing the stance toward decision making alters the causal factors and, I suggest, helps make what is now a rarity more common. But my immediate concern is that the bias be noted, by both those in positions of political power and those not, so that it may be guarded against in the case of waste where the solutions, more likely than not, need to be for the long term.

Self-Inoculation

The difficulty of being in a position of responsibility is that one is responsible, but it is an axiom of politics that those responsible should so arrange matters that they may take responsibility but not blame.

Chairs of committees thus arrange subcommittees to make recommendations regarding sensitive matters. Administrative heads appoint aides to handle hot issues, taking credit for the choice ("the best we could find"), but leaving themselves room for insulation from the actual decisions if necessary ("under the circumstances").

Legislators and other politicians might seem unable to inoculate themselves in this way. After all, they must vote or risk public exposure if they do not, and in voting they cannot hide behind others. It is they, not their aides, who vote. But axioms are axioms, and those in political power are subject to them no less than anyone else, especially in regard to decisions on the environment.

I sketched in Chapter 1 the alternatives that face an elected official who needs to take cognizance of a problem but does not want his or her ox gored:

1. So decide that everyone's interests are met.
2. So decide that, even though not everyone's interests are met, it can argued that the public interest is met.
3. Do not decide at all, and let experts make the decision.

In each of these cases, an elected official, political appointee, or bureaucrat is immunizing himself or herself from the bad effects of any decision. The political task is to so arrange matters when decisions must be made that the first and third alternatives occur most often and that when the second does occur, one can make a powerful appeal to the public good to justify denying some interests.

The methods by which elected officials so arrange matters as to inoculate themselves are well-known—sending to committees "for further study" matters that are politically too sensitive, compromising on key parts of proposals to placate various interest groups, trading votes, especially among those of different political outlooks, to create greater majorities so that decisions will be seen as mandated by a variety of political forces, an obvious compromise in the public good, and so on.

The decision by the Environmental Protection Agency (EPA) to drop a proposed requirement that those operating incinerators recycle one-fourth of the incoming waste may be an example of such inoculation. The decision was made by the EPA under pressure from the White House and, in particular, a panel headed by then Vice-President Quayle.[42] Certainly one way to read the result is that a supposedly independent regulatory agency, the EPA, took the political heat for a White House decision, immunizing the politicians involved.

It is particularly likely that decisions regarding waste will be made in a way, if possible, to immunize the politicians who must make them. For these decisions will of necessity harm particular interests of particular voters. Unlike building a factory or creating a public park, where some in an area to be affected will gain and some will lose, virtually no one who will be immediately affected by waste can see any benefits at all, except those common to all, such as having their garbage and trash removed. For a politician to decide, for instance, where an incineration plant ought to go is an instant guarantee that some percentage of voters will be upset. It is a political reality that no one wants incinerators in their own backyard, and no politician acting as a politician, to remember Hume's admonition, is likely to ask some constituents to sacrifice their own well-being for the public good when the well-being of so many others could be sacrificed just as readily.[43]

Politicians, in short, will be reluctant to make any choice regarding waste management that will harm or will be perceived to harm the interests of any constituents. Even trying to solve waste problems by getting rid of the waste, shipping it elsewhere, is not without its difficulties. New York now loads garbage into boxcars for shipment to the Midwest, but choosing *where* to load the garbage is as fraught with political peril as it would be to choose an incineration site in midtown Manhattan. The claim was that the operation would be neat and tidy because what would be loaded into the boxcars would already be baled, but as one resident in Long Island City says about a "transfer station" for garbage, "On paper it sounds sweet—no odors, no rats, no problems. . . . It never works out that way. Forklifts will puncture the bales."[44]

That such decisions regarding waste will of necessity harm the interests of some constituents is not the only reason politicians will want not to have to make them. They have another reason that makes them look less self-interested—and less political. Any public policy decision about the environment and waste management in particular requires understanding some very complex science. Such a decision may seem pre-eminently suited for experts, who can pronounce the names of the chemicals[45] and are best positioned to know which chemicals are toxic and which are not, which form compounds and which do not, which leach quickly through what kinds of soil, and so on. The general tendency, quite understandably, is for those in political power to hand over the decision making to supposed experts in the area.

This deference to authority seems so natural under the circumstances that it may seem inappropriate to mark it—as though it were a feature of decision making to guard against. But it is exactly that.

First, large areas of the field have not been thoroughly researched, and for anyone to assume thorough knowledge here would be a mistake. In addition, as we have noted, areas that have been researched are subject to error: Experts may not know what they think they know. Even worse, the huge increase in scientific research over the past 75 years or so has not only expanded our knowledge but also "expanded the universe about which we cannot speak with confidence. . . . Our techniques for finding new dangers have run ahead of our ability to discriminate among them."[46] We need only think of the huge number of new chemical compounds created each year and used without testing for their hazardous effects. Polybrominated biphenyl (PBB) is a sterling example. We know how to make such compounds but "cannot speak with confidence" about their effects.

Second, although we may be inclined to think that value judgments slip in only to fill the gaps left by ignorance, in fact, even with full information, reasonable people may disagree about a variety of matters, from whether certain risks are worth taking to whether certain expenditures made up front are better than uncertain, but probably greater, expenditures made later. Even if everyone agrees, for example, that a particular situation represents a gamble and that the odds are long or short, they still may differ as to whether to be cautious bettors and protect whatever gains they may make or be high fliers and go for high stakes. We may choose to sink all of our expendable capital into a waste-water treatment plant, in the hope that will solve all our problems, or we may purchase less than perfect machinery, judging that technology will develop quickly and that the funds may be better spent diminishing waste at its source. In short, even with full information, differing values may affect the very way the problem is conceived, the assessments of risk, and the selection of decision-procedures to respond to risk.

Third, even if one were to disagree that value judgments can affect us even with all relevant information, we will never know that we have all relevant information. Indeed, the effect of politicians trying to inoculate themselves from harm by handing over decisions to scientists is that scientists enter into a black hole of claim and counterclaim and politicians find themselves hard-pressed not to slip in there with them. It is a black hole because everything goes in and little comes out that can be acceptable to the scientific community as a whole until the problem has moved so far along its trajectory that any harm caused will be that much more difficult to mitigate.

The vision of decision making that requires for us to reach a rational decision that we have all relevant information compels scientists to search for more relevant information about whatever is at issue: That is their job, we ought to say. It is what they, as professionals, are obligated to do. But no rational decision can be based on anything other than truth on this view, and of all the propositions that might be true, only those that make it through the winnowing process of the scientific method are acceptable. The objectivity of any decision about waste and, more generally, the environment is supposedly ensured not just by having politicians, who are presumed to be biased, defer to scientists who are presumed unbiased because they have no political stake in the outcome of their researches, but also by having claims that matter to that decision vetted by the scientific process. They must be reproducible by other competent scientists, and, if they pass through the process, they are confirmed, and that is their mark of objectivity.[47]

But the effect of this process on decision making is that it never ends—for two different reasons. It *cannot* end because there is never an end to the quest for all relevant information. That is how it should be, of course. I am not complaining about this feature of the vision, as it applies to science, but marking out its implications for decision making. The quest also never ends because the information is always incomplete, by this vision. Recommendations must be made because decisions must be made, but those recommendations are always subject to second quessing. The best that can emerge is a slice, at some particular time, of the unending *process* that scientists are engaged in. Any report of the state of things at that slice of time will necessarily be contentious, for it will fail to anticipate the new discoveries that will be made even while the original report is being digested and will fail to be timely because the political process itself takes time and guarantees that the report will be viewed as outdated by the time it is considered. That is one reason it will always be relevant to play the epistemological card. Any scientific report that suggests a policy will necessarily be contentious, because, in addition to the scientific propositions, are other propositions that are not scientific, that are value judgments about how we ought to weigh risk, how we ought to judge when we lack some information, and so on.

We will have many gaps in our knowledge, and value judgments certainly slip in there. Because we do not know, for example, whether PBB causes harm to humans but do know that it causes harm to animals, should we ban consumption of livestock with PBB? With or without all relevant information, we cannot decide such a question without making some assessment of the value of human life over the value of the banned livestock, damage to the farm economy, and so on.

Such a judgment can be made by an expert in chemistry, for example, but it ought to be recognized that it can equally well be made by an elected official—and ought to be. There is nothing pre-eminently scientific about such judgments, and although a proper judgment will require some mastery of scientific knowledge, such mastery is not itself sufficient. Such judgments are pre-eminently political, dispersing benefits and burdens and risks among citizens and interest groups, and their status does not change if we note that a proper judgment must take cognizance of information best obtained from scientists.

The expert here is like a dentist making recommendations to a patient, and the patient must make the final judgment. To let the dentist decide what a patient can afford as well as determine the state of one's teeth is simply to allow an expert in one area carte blanche over areas not a matter of his or her expertise. A dentist is no more an expert in what a patient can afford than a scientist is an expert in the value choices endemic to public policy decisions.

Fourth, in the case of waste, at least, choosing an expert requires an expert. To put the matter less paradoxically, although some persons may well be objective judges about what ought to be done and neutral between alternative solutions, we are not in any position to know who that person or those persons might be. A great deal that looks objective in science turns out not to be based on any evidence about actual harm but, as with PBB, on the limits of detection by our instrumentation. That is an objective measure, certainly, but not one essentially connected, as the figure might imply, with harm or health. Similarly here, an objective basis may exist for making a choice among possible scientific advisors—like the total cost involved—but since there is little agreement among scientists as to what to do and whom to ask, it is too much to demand of those without scientific training that they pick and choose. Picking someone with enough scientific training and letting that person do the choosing does not solve the problem primarily because there is no neutral territory in the field. In short, even if we wanted to let an expert decide, we would not be able to.

The effect of these points about experts is that we should look with skepticism at an argument that prima facie seems a very powerful one. For just as we would be aghast at a dentist making judgments where he or she lacked expertise, so it is no surprise for politicians to tell us that we can hardly expect them to make scientific decisions: "You want *me* to make a judgment I am wholly unqualified to make?" The thrust of this query is that certain areas of competence require judgments by those competent in the area—issues of law by those trained in the law, of medicine by those trained in medicine, and so on. Just so, a scientific issue ought to be settled by scientists on this view, and a right-minded politician, aware of his or her own lack of expertise, ought to turn effective decision making over to those qualified by training and experience to understand the issues and to make an objective evaluation.

This seems a compelling argument. It has as extraordinary and tenacious a hold on the minds of those charged with making decisions about these matters as the view that they cannot make a fully rational decision without all relevant information. Yet, just as the latter has political benefits—by encouraging inaction, for example—so does the former—by making sure that no politician's ox is gored by any decision because it is made on the recommendation of those who supposedly know what ought to be done.

It might be that self-interest and the public good coincide at this point, but, in fact, they do not. We are *all* in the situation just described. Scientists are no better off than the rest of us in facing a situation in which there is a great deal of which we are ignorant, in which what we think we know we may not know, and in which we must act cautiously, given the harm that may occur from our present acts and omissions.

Scientists can presumably understand the scientific issues better than someone without scientific training, but that competency to comprehend part of what is at issue does not create a competence to comprehend all that *ought* to be happening, to make judgments about how tax monies ought to be spent or what sorts of capital investments are necessary and sufficient. These judgments are pre-eminently political, and because they are inextricably linked to any supposed solution to the problem of waste, for instance, that problem will have to be solved by those charged in our society with making such public policy decisions.

All this may thus seem depressing. Those who will have to make decisions about waste management are just those who operate within a decision-procedure that encourages them not to decide and then, when they have no choice but to decide, skews the choices available to them by transforming everything into interests, ignoring those problems in which not enough interest has been shown and then tending to fasten on solutions in which the bad effects are far removed and the good effects occur before the next election. A large class of environmental issues become problems only gradually, and the most obvious sorts of solutions seem to have heavy up-front costs with benefits only long down the road. Therefore, that those who must make decisions about the environment use this mode of decision-procedure does not bode well for environmental issues. But we can at least change our stance about how to make decisions in situations of uncertainty and so change one causal factor within that decision-procedure.[48] In any event, self-knowledge will allow us to guard against some of the most pervasive, and thus least noticed, features of the system we are in, and it is a good first step in coming to grips with any other relevant causal factors.

Who Decides What?

We have been focusing on the process of decision making, on how any issue that gets into that process is transformed and its solution skewed by the process itself. But we need to look at the issues and, in particular, at what sorts of issues get into the political process and at where they make their entry. One disadvantage of being a political official is that one faces all sorts of problems, some of which one is simply unable to solve.

The easiest way to appreciate this point in terms of the environment and of waste in particular is to look at the scope of issues. It is almost useless for a municipality to pass a bottle law requiring that all bottles within its limits carry a deposit. The act can at best have symbolic significance. For the City Council of Irvine, California to declare Irvine a clean-air zone would be like Berlin declaring itself a nuclear-free zone. If a muncipality passes a bottle law, bottlers will refuse to relabel their bottles for so few customers, and the ordinance will either pull in empties like a magnet or exclude certain sellers from its borders, depending on its details. Bottle laws in Oregon and Michigan, to cite just two states, have been effective because the states encompass large enough areas and have large enough populations to preclude such problems. Similarly, it is almost useless for

Canada to pass legislation about acid rain because the bulk of the problem comes from the United States. What is needed is an international solution.

One can instantly multiply examples. Even though 2 percent of municipal waste comes from disposable diapers, a municipality is an inappropriate political body to require that those within its confines use only biodegradable or cloth diapers.[49] Demanding that of its residents would be like demanding that they purchase all their milk in disposable and recyclable plastic pouches. The residents will simply not be able to find such commodities, and the municipality would not be able to enforce its regulations in any event.

The difficulty with the scope of regulations is a standard problem in any federal system: Some problems transcend state boundaries. So one traditional justification for federal action in the United States has been the intractability of some problems to solution by individual states. The same sort of issue can arise between nations, as with acid rain or dumping waste in the ocean, and between municipalities or other political bodies within any state. The scope of an issue may be too broad for the political body considering it.[50]

The difficulty can run the other direction as well. Nothing might seem more appropriate to local control than setting fees at the local dump. Yet, high fees at a dump, set to cover the high costs of trying to contain waste or to discourage local dumping, will drive up costs at other nearby dumps, which would soon be inundated otherwise, and all that can have international repercussions as various firms soon discover the profit in moving trash and garbage long distances to locations where dumping fees are significantly lower—Cornwall, for instance, or African nations such as Nigeria where the charge can be as little as $2.50 per ton for toxic material that would cost $3,000 to incinerate in Europe.[51]

The issues that come before a political body do not necessarily fit the competence of that body. The proper solution may require a larger scale, and a local solution may create problems elsewhere and produce, overall, a net loss.

In short, there is no guarantee at all that the right person or political body will be faced with the right questions. A local county council will find itself having to decide about the appropriate size of a proposed landfill when it knows that a good percentage of the fill could be handled with a state bottling law that would effectively remove bottles from waste. It needs to sort through its problems as well as its trash and make the time-honored political choice: Pass the buck.

Of course, that is another way for politicians to decide not to decide an issue. They can decide that it is not within their *political* competence and either pass it on to another political body, for example, a state legislature, or to other sorts of political bodies, such as the court system. One of the virtues of the federal system is that there are many sources of political power: None are supposed to so dominate the others as to become so powerful, for instance, as to put liberty at risk. But that virtue can be a fault because with so many other sources of political power, it is easier for one to plead incompetence. The plea will be plausible in regard to waste because it is one of its more frustrating features that the source of the difficulty for a particular problem tends to require a broader scope for its control than can be covered by the political bodies before whom the problem usually appears.

On the other hand, this general difficulty of fitting the problem to the right political body can be used advantageously to make distinctions between kinds of waste and to see how to get a handle on some of it. I suggested that if the reader has any image of waste in mind while reading this, I would prefer it be the sort of trash and garbage we all have in our own households rather than, say, the hazardous waste of industries. The reason for that, I suggested, is that industrial waste presents a cleaner problem. Because it comes from industries, it can be identified and isolated relatively easily, and the political bodies that must exercise their competence regarding it are identifiable as well.

Germany has taken advantage of the capacity to identify and regulate businesses and industries to require that they "collect and recycle the cans, bottles, cardboard, paper and plastic used to package their products."[52] It might be easy for a homeowner to deny responsibility for cans found dumped on the lawn, but it is hard for the company whose logo is emblazoned on the cans to deny that it used them to sell its product.[53]

In the United States, the Emergency Planning and Right to Know Act puts citizens in a position to know what sorts of pollutants, and how much, are being produced by manufacturers.[54] Although the Act has primarily been used so far to give citizens sufficient information to pressure heavy polluters to cut back on dangerous or questionable emissions, the information gained can be used for other purposes as well. California will begin to monitor smokestacks and other major sources of pollution "with sensors or tracking devices to measure emissions on a continuing basis."[55] The aim is to provide companies with the information they need to comply with new standards on emission. No particular method of complying is required, and "companies that cannot meet the required reductions, or choose not to, will be able to purchase the right to pollute from companies that exceed the requirements."[56]

Sources of waste and pollution are not all equal, and we can use what prepackaging already occurs in the sources from which pollution comes to begin to control it. But we must marry the appropriate political body to the proper source. In the case of industries, what is required is the political will of those political bodies regulating industries, namely, states, nations and, where the industries operate beyond national boundaries, the appropriate international bodies. The problem is most urgent regarding hazardous materials, and what is required is that those producing hazardous material as waste identify it, seal it effectively, label it properly, and transport it safely to incineration plants that they may be specially taxed to build and support.[57] But it will not take just the political will of particular states or nations. If one nation requires industries to cleanse themselves, then the industries of that nation are at a competitive disadvantage with the industries of other nations that are getting a free ride by not having to pay for the disposal of their toxic wastes. So when the problem is international, what is also required is an international effort to ensure that the restrictions imposed are uniform from nation to nation and take effect simultaneously. Just as the United States has the capacity to act effectively to provide such support for state regulations within its borders, so the Common Market could act to do the same for the nations of Europe.

These features of industrial hazardous waste—that it is clear enough what needs to be done to clean it up and clear enough who is politically responsible—make it a more tractable problem than the waste of our trash cans. I do not mean that the problem is easy to solve or even will be solved. The cost is high, and the political cooperation required enormous. Yet, it is clearer what can be done about one aspect of the problem.

A major difficulty with the waste of our trash cans, on the other hand, is that it does not present such a clear target for action and that those who must make decisions about what to do, like those on a local county commission or a city board charged with waste management, face many problems that are beyond their political capacity. Trying to solve such a problem will sometimes only aggravate it. New York City created enormous difficulties for itself and other areas of the country when it raised its tipping fees from $18.50 to $40 a cubic yard. As we saw, business went elsewhere. So the City lost money it needed to maintain its dump sites, and other areas of the country were inundated with trash that would have quickly filled their dump sites without such remedial action as increasing their fees, thereby tending to drive those who used their dump sites farther abroad.[58] Because of the limited scope of its political power, the body that opted to increase the tipping fee had limited options, and so what would presumably have been the best choice to solve the original problems was apparently not a politically possible choice. Or, turning the viewpoint around, what looked to be the politically best choice for New

York City turned out not to be the best choice, all things considered. The ontoward effects of the decision is evidence of that.

The danger of encouraging political groups to pass along such problems is that everyone may pass the political buck, but political bodies ought to be encouraged to do what can be done effectively by them and pressure other political bodies to take on the broader measures that remain. Local government could require significant deposits on any automotive and truck batteries purchased within its confines, and although that would drive purchasers elsewhere, it would also produce a new group of lobbyists who tend to be influential, namely, the economic interests affected by the loss of business. Such an act by a local governmental authority would thus act to discourage in the long term the dumping of batteries within its environs and also help in passing on to a broader governmental unit, perhaps the state government, the broader issue that can only be handled effectively at a higher level.

Whether local bodies can act decisively is another issue—as is the issue of whether they ought to act decisively. As the example of New York City's raising its dumping fees shows, local action can have quite adverse consequences, both locally and nationally, and there is a real question whether the best interests of all those involved, local residents and others as well, are best met by local responses.[59] But it is also an issue whether local responses are possible. The overlying federal structure of our system and the recent flurry of legislative activity at the federal level, combined with "a confused and delayed regulatory issuance and implementation process," have made it always, it seems, an issue whether a local jurisdiction has the legal capacity to respond.[60]

A well-designed policy would presumably sort out the conflicts that exist between various levels of government and allocate problems to those levels, and so to those political bodies, best able to respond to them without causing further problems. But we now begin to press toward an unrealistic and naive idealism. Our political system is itself, as I have remarked, a natural social artifact that is immensely complex, and to suppose that somehow, to resolve environmental issues, we first need to change the structural features of our political system is to condemn ourselves to doing nothing.

We now, however, begin to approach the pressing question, "But what are we to *do*?" To respond to that, we need to examine our options and thus, I argue, the way we tend to conceive of our problems.

CHAPTER FIVE

The Options

It seems that since the 1980s we have made enormous strides toward recycling. Cities have invested large sums in special trucks; citizens haul refuse to the curb, having neatly separated cans and bottles and plastics from newspapers and newspapers from other refuse; one now routinely finds products marked recyclable or made from recycled products. We seem on the crest of a wave where recycling will become the dominant force in the way we manage our waste.

The figures might seem to bear out this assessment. In 1966, we recovered 0.18 pounds of the 2.66 pounds each person in the United States produced, on average, in waste. That is a 6.7-percent recovery rate. By 1988, the rate had increased to 13.1 percent, with 0.52 pounds recovered for each person, on average, each day. We thus almost doubled the rate of recovery, but we are also producing more waste—2.66 pounds per person in 1966 and 4 pounds in 1988, 87.8 million tons per year versus 179.6 tons per year. So the gain in the rate of recovery was more than offset by the increase in the rate of production of waste. Whereas we somehow disposed of 2.48 pounds of waste per person in 1966, we had to dispose of 3.48 pounds per person in 1988.[1]

Although we have increased, and continue to increase, what we recycle, we also have increased even more the waste we produce. At the rate we are going, we will never catch up. It is that problem that provides one explanation why, in many current discussions about waste, incineration tends to emerge as the preferred option, always combined with a judicious mixture of landfilling and recycling, but dominating and driving the whole. As the president of the National Solid Waste Management Association, Allen Moore, says, "Waste combustion is now seen as a cost-effective and environmentally sound means of solid waste management for many communities."[2]

The increase in incineration capacity in the United States has thus been enormous since the 1970s. In 1970 only 1 percent of solid waste was incinerated; by 1992, nearly 20 percent was. The percentage of increase is thus significantly greater than the percentage of increase in recycling.

In 1989 there were 172 incineration plants, 122 of which convert waste to energy. "In addition, there were 31 under construction and 74 in the planning stages."[3] It is projected that "from 1990 to 2010, there will be a need for . . . 300 new plants burning municipal solid wastes."[4] Many of these are modeled after the massive incineration plant in Detroit that can burn 3,300 tons a day. The environmental cry of the 1990s, it appears, is "Burn, baby, burn!"

Yet, incineration is the preferred solution to a problem that our current practices create *because* of that practice: Its pre-eminence is an artifact of the natural social artifact that is our current waste management system. If we were not producing so much more waste, or producing so much that could not be recycled, for example, it would not seem such an attractive choice.

Consider, by way of analogy, the natural social artifact that is our transportation system. It is a system heavily reliant on private automobiles, made increasingly necessary by the far-flung suburbs and country homes they make possible and by the decreasing availability of convenient public transportation systems. If we were to ask how we might save energy in transportation, making automobiles more fuel-efficient would emerge as a preferred option. For from *within* that system, alternative forms of more energy-efficient transportation—light-rail systems, subways, buses—often seem too expensive for the few riders they would be able to attract and serve. Suburbs are already so far-flung, for one thing, that putting in a light-rail system with its need for fewer rather than more stops would require of most commuters extensive trips just to get to the rail stops. So increasing the fuel efficiency of automobiles would garner much greater savings much more quickly and with less disruption to our ongoing practices.

Yet, what looks plausible from within a natural social artifact often takes on a new cast when we examine the social artifact itself and ask, "Is it reasonable?" Faced with problems created by a natural social artifact and with constraints produced by it that make some options significantly less eligible than others, we may find it difficult to back off and examine the natural social artifact itself, to ask, "How did we get ourselves into a position where incineration emerges as the preferred option?" That is what we need to do, and when we do, we will discover that our real choice is not between incineration and other means of waste disposal but between fundamentally different ways of moving through this world.

Why Incineration?

New York City is trying to impose restrictions on how much salt can be used to de-ice roads in the Catskills, and it is difficult to understand how such an attempt could even begin to make sense without seeing it as a function of a pre-existing network of reservoirs and water use the City is trying to protect.[5] In short, what would seem otherwise implausible is made plausible by the natural social artifact already in place. The City's attempts to regulate building and land use in upstate New York is an artifact of its having an extensive reservoir system that must be protected to preserve the purity and taste of its water without expensive filtering.

In the case of waste, we have local political bodies such as city councils and county commissions faced with massive amounts of trash to dispose of. What is already there is not going to go away, and so far as such a political body can know, what is coming by the truckloads is not going to be reduced significantly any time soon. At least, such bodies cannot plan on any reduction but must instead plan on the basis of the usual projected increases. So whatever the long-term benefits of the attempts to reduce the total amount of waste produced through changes in packaging, for instance, waste comes as something to be managed immediately from the perspective of such local political bodies. From within our current natural social artifact of waste management, that is, where the amount of waste produced is an artifact of habits of use and production not likely to change any time soon, it seems eminently reasonable to consider what ought to be done with the increasing amount of trash coming in—instead of speculating on what might be done if such trash were not to come. Something must be done because leaving trash to accumulate is not a politically viable option. Any city administration, for instance, will take as much heat from trashing the collection of trash as it will for failing, for example, to remove ice or snow from the city's streets.

If we look at the problem from the point of view of such a political body as a county commission charged with getting rid of the waste the citizens of the county produce, we can most readily understand why incineration tends to emerge as the preferred solution, why it tends to dominate and drive discussions of waste management. For that perspective seems to necessitate that we conceive of the difficulty as that of managing waste. The problem *presents* itself to such political bodies as county commissions as a problem to manage: "What are we to do with all this stuff?" Waste comes as a given, and the only question is whether it should be buried, recycled, incinerated, or shipped somewhere else. The problem becomes that of getting rid of it in some way, and the criteria for an acceptable solution to this problem turn on the advantages and disadvantages of alternative methods of disposal, using such time-honored criteria as cost-effectiveness, ease of handling, and so on.

Landfilling has traditionally been the preferred option because, among other virtues, its short-term benefits have been high and its long-term disadvantages such as ground-water pollution so far removed and incalculable as to have a relatively negligible effect on decisions about siting. Landfilling best met the criteria for decision making that I have suggested tend to skew choices for those who are charged in our society with making such decisions.[6] But for those political bodies, the calculation of short-term benefits and losses has changed just as old landfills are running out of space.[7] The histories of supposedly safe landfills have jeopardized claims that any landfill site will have a safe future,[8] and increasing land use has made finding a landfill site more difficult and the political price of any siting high. Those who purchased land and built in one part of a county, far removed from the county landfill, are a potent political force in any attempts to site a new landfill.

In addition, as the total amount of trash created increases, any new landfill will fill more and more quickly, as its opponents will note, and the problem of what to do with the trash will return quickly even if a commission could find an acceptable site for a landfill. It will return with a vengeance because the likelihood of finding yet another politically acceptable alternative landfill site will be small in a county with, say, two already.

Landfilling is thus no longer the preferred option. The political price of siting a new landfill is so very high that a political body such as a county commission hardly dare make the attempt any more. As such political bodies have scrambled for solutions that are cost-effective and have an immediate impact on getting rid of the waste that accumulates and will continue to accumulate whatever the problems they face, one alternative has been to ship the waste elsewhere. "Out of sight, out of mind" is a political maxim with enormous appeal in such a situation.[9] But shipping the waste of a county or a city elsewhere is becoming ever more expensive as various other landfills reach their capacity. Waste must be sent farther and farther away as other local jurisdictions either discourage dumping with high fees[10] or prohibit entry from outside sources to preserve as much space for themselves for as long as they can and to protect themselves from being, and being perceived as being, merely a dumping ground for other people's trash.[11] So shipping trash elsewhere can, at best it seems, supplement other solutions. It cannot in itself function as the dominant solution. Clearly not *every* jurisdiction concerned to get rid of its trash can ship it elsewhere.[12]

Because the problem of what to do with trash will appear again relatively quickly if landfilling is chosen, it seems preferable to pay the political price for siting an incinerator to gain its advantages and to bypass as many of the disadvantages as possible that landfilling now entails. In short, incineration emerges as the preferred solution because, once waste is assumed as a given, it most readily meets the various criteria, political and otherwise, for managing such a problem.

One criterion is that the preferred option will solve the problem once and for all. This is an ideal, of course, and incineration comes closer to realizing it than landfilling. For landfilling guarantees that the problem of what to do with all the waste will reoccur as a problem. A landfill will fill up eventually, but an incineration plant can keep on operating through its lifetime as long as it is maintained.

Of course, an incineration plant may be overcome with more waste than it can handle so that other plants may have to be built. So building an incinerator does not prevent some form of the problem of siting from reappearing sometime in the future. But at least the old incinerator will not have to go out of service, unlike a filled landfill, and if the present one is made big enough, the need for another will be that much farther in the future—one reason for the huge size of these incinerators. And, of course, we must dispose of the ashes that incineration produces. But no one is suggesting that we can incinerate *everything*, without a residue. We can incinerate much, and, it is claimed, we have no reason now to suspect that incineration will produce any new problems in the way in which landfilling produced unanticipated problems with ground-water contamination, for instance. So incineration has the virtue of coming closer to resolving the problem once and for all than landfilling.

Incineration has another important virtue as well. Its hold over us as the solution rests in part on its being the only solution that leaves everything else pretty much as it is. If we opt for incineration as our dominant way of managing trash, we do not need to change anything except where the trucks go when they haul our trash. We simply build incinerators instead of finding waste dump sites. It is, in short, the only solution that leaves our ongoing natural social artifact of waste management effectively in place. It thus satisfies the following condition: The preferred option requires the least changes in our ongoing practices.

"Once is enough!" is a political principle that plays a crucial role in making public policy decisions, including those about the environment and waste. The expenditure of time and energy and political capital is too high to make it worthwhile going through the process of making such decisions again. So it would be best if the solution regarding waste were the one most likely to mesh with our ongoing natural social artifact—and so the one most likely to work. Landfilling does that, but incineration does it without the long-term bad effects of landfilling. In addition, incineration has the clear advantage over recycling because it requires no fundamental change at all. We can simply leave everything as it is. We need not change what we produce or how we produce it. We just take our trash and burn it instead of burying it.

So we have none of the expenses that would be tied to creating a new social artifact—none of the concerns about building recycling centers, purchasing a fleet of specially designed trucks to collect household trash,[13] shipping sorted trash wherever it must (and can) go,[14] or convincing taxpayers that the costs of recycling are worth the benefits.[15]

But we not only have none of the expenses tied to creating a new social artifact, we have income created by incinerating our trash. We not only do not lose tax money, we gain it. On the one hand, we have none of the expenses created by having to put aside land for a dump site—the loss of tax revenue while it is a dump and, presumably, for some long time after, the expenses associated with maintaining the dump site long after it has ceased to be useful in order to ensure that there is no groundwater contamination, and so on. On the other hand, there is revenue produced by selling the energy created by burning the trash. It would take a politically otiose county commission to decide not to save taxpayers the money created by turning their trash into electrical energy that can be sold. Whatever those profits may be, whether they will be more than enough to cover the costs of building and operating the incinerator, or whether they will just help the taxpayers in some measure bear those costs, they are more financial help than can be expected from any other source. A county commission thus appears in the enviable position, in choosing to incinerate, of satisfying the following condition: The preferred option will cost little.

After all, any money put into waste is wasted. Waste is not a product, and so, by this line of thinking, money spent on it is itself wasted unless it can somehow be recovered through a product, such as energy, produced by the waste. The same sort of reasoning has persuaded many municipalities to convert their sludge from sewage treatment plants into fertilizer. There is a double savings: There is no need to spend money to do anything with the sludge, such as shipping it elsewhere, and there is the income generated by selling the converted sludge.

Landfilling has the apparent virtues of being as inexpensive as the land and the hauling costs, but the waste that is buried gains no one a profit, the land is lost for any other use for some time, if not all time, and the leaks that have plagued landfills are expensive to monitor for and, if found, to contain and clean up. But we can recover at least some of the funds put into incineration, it is claimed, through the creation of electrical energy and the savings engendered by not having to look after a landfill long after its useful life is gone. So the waste is not a complete financial loss, as it is with landfilling, and the only new funding required is that of building an incinerator that will pay for at least a good part of itself. Recycling can also save money, of course, as it does in the recycling of aluminimum, for instance, but it requires time-consuming and thus expensive sorting and handling that is not well-suited even for many products that could be recycled.

A more comprehensive and detailed analysis of the relative costs would not show much more than what this rough comparison shows, I would argue, because few things are clearer regarding projected costs than that they are always inaccurate.[16] The best we can get are rough comparisons, and on the basis of the sort of rough comparison any political body such as a county commission is going to be able to make, incineration wins out. This is particularly true as building an incinerator is something it can *do*, whereas it seems impossible that any county commission can change an existing natural social artifact—the habits of use of all the citizens of the county, reinforced by a massive commercial system that makes some environmentally sensible choices impossible.[17]

Besides, incineration takes care of almost everything at once. We do not have such residual problems as what to do with what cannot be recycled. In short, incineration comes close to satisfying another criterion for a solution: The preferred option will handle all the waste at once. We do not want a solution that creates another, more difficult problem of disposal. How could we dispose of what our preferred method of disposal could not handle?

Landfilling satisfies this condition readily because it need not be supplemented by anything. It (supposedly) leaves no residue of a problem. That it seems to satisfy this condition has much to do with explaining why it has been for so long the method of choice for managing waste. Incineration produces residual ashes, including heavy metals that cannot be incinerated, but, it is argued, these can be retained in the incinerator, are gathered relatively easily, and would take up far less space to bury than would all the trash that will be burnt. Unfortunately, that residue is concentrated and far more toxic in the form in which it must be handled than it was when it entered the trash stream. This criterion thus creates the biggest problem for incineration. It itself produces a waste product that, arguably, must be given special and expensive handling (at $210 per ton as hazardous waste versus $23 per ton in an ordinary landfill for the city of Chicago).[18]

Some see this as an advantage. At least, it may be argued, the toxic elements are concentrated and so *can* be given special handling rather than being dispersed widely in a landfill. In short, we know how to supplement incineration, and incineration does have one big advantage: What goes up in smoke disappears, at least as far as those who run an incinerator are concerned. Landfills have been known to seep, and then what we have thrown away returns. And if landfilling is no longer an option on other grounds, then, by this criterion, incineration is the preferred choice, despite its not handling all our waste at once.

For we not only know how it can be supplemented, we also know that it need not be supplemented with anything else. It not only piggybacks on our current natural social artifact of waste management, it also takes advantage of existing technology. Recycling poses problems because we need to change the way in which we package many of our products to make them recoverable, let alone recoverable in an economically efficient manner. Imposing such changes is certainly beyond the capacities of any county commission. The technology for incineration already exists, and so the option of incineration presents itself to a county commission as a ready choice, requiring nothing new, requiring no promises that future development may or may not bring. Incineration thus satisfies the following condition: The best solution will not require any additional research but will take advantage of existing technology. Incineration arguably has a track record, in short, and just as we can piggyback on our current practice, so we can simply extend our current technology.[19]

Thus, if we construe the problem as it presents itself to us, as waste to be managed, and if we must choose between recycling, incineration, and landfilling, incineration looks like the winner. It has more advantages and fewer disadvantages than recycling or landfilling.[20] It is especially likely to be chosen given the ways in which the political process tends to skew the choices. For, among other things, it presents no worse siting problem than landfilling, so that its immediate burdens are no worse, and it has immediate paybacks in the form of revenue.

We need not, according to this view, choose between the three options, as though picking one excluded the others. We can make incineration the primary mode of disposal and combine it with a judicious use of recycling and landfilling. What cannot be incinerated is either recovered through scrubbing and then buried, or it is sorted out ahead of time where possible and where that solution pays for itself through recycling. Incineration would be the dominant mode of disposal. We could recycle what could not be incinerated where that would pay, and we would bury what could not be incinerated when recycling would not pay or was not feasible.

The total package seems to overcome the objections to all the other proffered solutions. Incineration will, it is claimed, solve the problem of what to do with our trash once and for all, with the least changes in our current natural social artifact of waste management, with no need for any additional delay while further research is conducted, and with all the waste being handled relatively easily, with what little waste is left over being landfilled. In addition, and most compelling for some, incineration has a clear economic advantage. Besides the savings that comes from not having to worry about landfilling and its effects, trash is turned from a by-product of our way of moving through this world into a product through the production of electrical energy. It turns waste into money. The argument is not quite that we are getting something for nothing, but close enough for political purposes. Once the economic arguments are made, it is difficult for any political body charged with ridding a community of waste to resist them. To paraphrase David Hume yet again, why would a politician "persevere in a measure so disagreeable to all" his or her constituents as to tax them for getting rid of trash when incineration would effectively *lower* taxes?[21]

One objection to this conclusion that incineration is the preferred alternative is that in practice the idea turns out to be significantly less compelling than it appears in theory. According to John Dieffenbacher-Krall, an organizer for the Maine People's Alliance, "It was seen as a wonderful idea that would solve two public policy problems at the same time: the energy crisis and the trash crisis." But the incinerators in Maine have been hit by operational problems, complaints about the environmental effects, and legal battles. The Maine Energy Recovery Company, which operates a $103 million incinerator in Biddeford, Maine, reported losses of $500,000 a month at the end of 1990 and is asking for an increase in dumping fees to $47 a ton.[22] The most charitable reading of the problem is that, like many public projects, from subways to new airports, the actual costs and problems are always significantly greater than projected and the benefits achieved are always fewer.[23]

In addition, an obvious problem with incineration is that it will itself produce a demand for waste. If incineration is to produce a profit, the plant must have raw material to incinerate, and so one of the compelling reasons for building an incinerator—turning waste into profit—will discourage both the reduction of waste and attempts to produce more recyclable products. Put more starkly, to the extent that incineration becomes the dominant mode of waste management, waste reduction must become a secondary consideration. We shall otherwise find ourselves with huge capital investments that are a burden to the taxpayers, unable to turn a profit because they are starved for waste.

But after presenting why incineration is the preferred choice of a political body such as a county commission and, indeed, ought to be the preferred choice given our current natural social artifact of waste management, it would be easy for us to get caught up in disputing its details, ferreting out what we could about costs, and comparing projected costs of one alternative with projected costs of another. Arguments can be made on either side, and consensus is difficult to achieve.[24]

Yet, as I have said, it is the way the problem presents itself to us that makes incineration emerge as the preferred solution, and we should not stay within the natural social artifact from which incineration emerges. We should not assume that our task is to take a given, waste, and manage it somehow. Yet, in opting for incineration as the solution to the problem of waste, we opt to maintain our current natural social artifact, for a particular way of moving through this world. If we think of the problem as external to us and our way of living, then we see ourselves as *facing* a problem rather than being part of it. It is as though we have already decided that we are going to live a certain kind of life, like the one we have been living, and just need to determine a way to clean up after ourselves.

Focusing on the Long Term

The proper frame of mind here, the appropriate attitude to have, is one that requires ignoring the immediacy of doing something with what we continue to produce, that fastens an eye on the long term, and that asks of the entire process, from our creation of waste to our disposal of it, how it can best be managed. What needs management is not the waste itself, for that is a by-product of our mode of living. We ought to ask how we can best manage our living so as to produce by-products that can be managed best.

Our real choice is thus not between recycling, incineration, and landfilling. That these are our perceived choices, and that incineration stands out from among them as the preferred option, are artifacts of the way the problem comes to us—as waste to be managed. Our real choice is between alternative ways of moving through the world, between, that is, alternative ways of living in this world. We now create in the United States one and a half to two tons of waste per capita per year. From disposable diapers to the disposable medicine containers of our old age, we move through the world leaving not only our natural residues but also an accumulation of material none of us could heft, visible trails that we may hardly notice except when backpacking or cleaning our yards of the litter of others, or, in an abstract way, when we try to determine how we each could create so much waste each year. If we fail to back off and look at our practice of producing so much waste that cannot readily be recycled, if we leave our style of moving through this world alone, that is, incineration seems the preferred alternative. But our style of moving through this world is not inevitable. We can change it by emphasizing waste reduction and recycling. If we cut the amount of waste that we produce to begin with, and if we learn to make the waste we must produce more amenable to recycling, then we will have a solution with more advantages and fewer disadvantages than incineration. But to make waste reduction and recycling our primary concerns, we must come to manage our lives.

That may seem an overwhelming thought. We have a difficult enough time managing waste, let alone *managing our lives* as well. Yet, if we are to have a practical long-term solution, we will end up effectively changing the way we live, and we ought to recognize that—up front. To realize that our habits of use are part of the problem, to see ourselves as integral to the issue rather than have us facing an issue external to us, is crucial if we are to begin to get a handle on the problem, in the sense both of understanding its full complexities and of working out a practical and effective set of solutions.

It is less abstract and overwhelming if we think of asking what variables are relevant. When we talk of waste management, and so tend to treat waste as a given, we find ourselves faced with having to rid ourselves of it by incineration, recycling, or dumping, rendering it less obtrusive by either changing its form or covering it up. But talk of waste *management* tends to obscure the creation of waste. We produce more than we need to, and we produce too much that cannot readily be recycled. Those are variables too, and once recognized as variables, we can make changing them dominant in our way of moving through the world. We ought to make dominant the following principles:

1. Reduce the production of waste.
2. Recycle what we can.
3. Reconfigure what we do produce to make it more amenable to recycling.

We will reduce up front the problem we face in managing waste and ease the managing by making more of that waste more appropriate for recycling. In the long run, the reconfiguration of what we do produce so that it is more amenable to recycling ought to come to dominate the whole. The economic incentives for doing that are already enormous for much that we produce, aluminum being the clearest example, and it should not be difficult to produce more incentives for other products.[25]

Our options here are not incineration or landfill, or a combination of both, or a combination of both with recycling, or of both with recycling and reduction. Choosing one will not preclude our using the others, and, with the waste we now have, it is difficult to conceive how we could not somehow use all these alternatives. But as we think of long-term solutions to the problems we face, we must realize that some option must come to dominate. As it does, it will fundamentally alter the way we move through this world and the world as well. Our real object of choice is thus some one alternative as the dominant mode along with an attendant mode of moving through this world.

Changing Habits and Practical Problems

There is no doubt that reducing the total amount of waste each of us creates each year would require changes in habits of use difficult to achieve. For example, parents now use an enormous number of disposable diapers to handle the waste that infants produce. If there were evidence that disposable diapers were associated with a higher incidence of cancer, one would not be able to give them away. If they caused severe rashes, that would no doubt cause most parents to use cloth diapers. But disposable diapers are one of the wonders of the modern world—a convenient and relatively inexpensive means of handling what is otherwise a messy chore. Only a deeply unrealistic optimism about human nature could move any to think that a significant number of parents would be moved by environmental concerns to cease using disposable diapers.[26] That is the trouble with much waste we create: It is the result of modern conveniences that are, indeed, convenient. In short, our habits of use have a built-in inertia difficult to change because they are responsive to conveniences difficult to give up.[27]

Indeed, even if we wanted to change our personal habits, we have at present few options for some products. As I have remarked, milk is purchased in containers in supermarkets. Home delivery of milk in reusable containers is a thing of the past for most Americans, and the containers used in markets are often difficult to recycle and so designed as to be wasteful of resources. In short, our personal habits of use are often reflective of the options available to most of us, or the options that are available to us reflect our patterns of use and thus our preferences.

Changing habits is never easy, and when the habits of use and the practices we have are so reflective of such conveniences and so deeply embedded in the economic system, we may find this another source of concern about our capacity to come to grips with the problem. The way we have to change is so fundamental and systemic. It is the source of this concern, the major shift in how we ought to move through this world, that may be partially responsible for our tendency to think the problem one of waste management. If it were a management problem, it would be a practical problem, with some solution or other, however expensive. But if we are to manage our lives as well? That thought may seem overwhelming. Yet if we think of reduction and of creating products easy to recycle as simply new variables to take into consideration, the problems become practical—and solvable.

Indeed, we are already making strides in solving some of them. Anyone can now purchase batteries with little mercury in them, significantly easing the problem of disposing of a very toxic substance. It is a major problem faced by incinerators that mercury tends to escape scrubbers. As we have seen, the large incinerator in Detroit was closed for this reason after less than a year of operation. A major source of that mercury is in batteries, and so reducing the amount of mercury in batteries can thus significantly ease that problem. It is a mark of how much progress has been made that in older models mercury amounted to as much as 7 percent of the battery's total weight and that we may now purchase batteries where the mercury's weight is less than one-tenth of 1 percent of the total.[28]

Or consider packaging. In Canada, over 50 percent of the milk being sold is packaged in pouches. Those still need to be buried in landfills, but since packaging accounts for about 30 percent of all solid waste by weight and about 34 percent by volume, having thin packages that collapse when emptied significantly reduces both the weight and volume of material that must be buried. We can get some sense of the reduction and also of the economical advantages of switching to pouches by noting that full "pouches take up 40 percent less space in storage than cans and bottles" because they leave little empty air in any carton in which they are stored. We also get some sense of how difficult it is to change consumer habits by noting that few inroads have been made in the American market because, it is claimed, consumers in the United States "have become accustomed to cardboard cartons and rigid plastic bottles."[29]

Or consider the packaging used for such products as ketchup. To keep the contents from being spoiled, they need to be protected from oxygen, and the packages used for them are complex because the oxygen barrier needs to be protected both from the contents of the container and from the outside environment. The original squeezable plastic ketchup bottle had six layers, including two adhesive layers to bind the oxygen barrier to the polypropylene that formed the inner and outer layers. The adhesive made it difficult to separate the components for recycling, and polypropylene is not recycled commercially in any case. To solve these problems, at least one ketchup company has created a new bottle, equally squeezable, made of polyethylene terephthalate (PET) that is used in bottles of pop and can be recycled. Whether it will be recycled is another question, of course. As John F. Ruston, at the Environmental Defense Fund said, "They had a container that was unrecyclable in theory and practice. Now it is recyclable in theory."[30]

Another solution, still too expensive to be commercially competitive, is to coat the plastic with a thin layer of glass—"less than one-thousandth the diameter of a human hair"—that is impervious to oxygen but "does not significantly contaminate the plastic, which therefore can be recyled."[31] Such solutions, if they work for ketchup, will work for other food products, such as fruit juices, the containers of which are not recyclable and take up a great deal of landfill space.[32]

Again, improvements in products that make them easier to recycle must be matched with improvements in recycling, but the point is that once reduction of waste and the creation of products that are easier to recycle are seen as simply two additional variables, the problems become practical problems, and these have solutions—driven by economic as well as political incentives. In short, once we back off from our practice of producing waste and examine it, we can come to grips with the practical problems the practice produces.

All this has been said with a view to showing how we can change our focus of concern—from how to handle the waste we create to how to move through this world so that we produce waste we can handle. Often one of the impediments to the solution of a problem is that it is misconceived, and I have been suggesting that the reason incineration surfaces as the preferred solution to the problem of waste is that the problem is misconceived in a certain way, with waste as a given. Conceive it in a different way, pay attention to the long term, and reduction and recycling come to dominate and drive any solution.

But little has been said so far about whether we *ought* to conceive the problem in a new way. If our object of concern ought to be a set of practices with its set of consequences, we need to assess the *whole*. Do we want those consequences? Do we need those practices?

CHAPTER SIX

What Ought We to Do?

The short answer is that we ought at least to do that which least wrongs our rights and least harms our interests, but short answers are rarely helpful. What is needed is a clear sense of when it is morally appropriate and reasonable to act and when it is not. My aim is to provide a number of cases that are both clear enough paradigms to give us guidance and decisive enough to suggest the properly careful cast of mind with which we ought to respond to environmental issues and waste in particular.

I mean, in short, to argue for a change in the way in which we *reason* about such issues. Disputes about environmental issues tend to have a particular and compelling logic. Someone will play the epistemological card, quite appropriately and reasonably, it seems, to argue that more evidence needs to be given of real harm. Then, after a great deal of work has been expended to gather more evidence to support such a showing, it will be asked, "But how *likely* is it that there will be such harm?" The question seems reasonable, of course, because showing that real harm *may* occur does nothing to show that we should prevent what may cause it. Real harm *may* occur no matter what we do, and what is necessary, on this view, is to assess the *likelihood* of such harm occurring. Then, and only then, by this mode of proceeding, can we weigh the benefits and burdens and properly weigh the risks to make a reasonable decision based on the outcome of that assessment.

However reasonable this series of moves seems, it can be frustrating for anyone concerned to bring about change. For it has the effect of reinforcing whatever may be the existing configuration of interests and expenditures in the natural social artifact in which it figures. Because it will frustrate anything that might change the status quo, this series of moves can be as much a burden to those who wish to build incinerators, for instance, or site landfills, as to those concerned about how our current natural social artifact of waste management harms the environment or how emissions of various gases may cause a greenhouse effect. But it may seem peculiarly frustrating to those who wish to ensure that harm is not done to the environment. For although they usually do not have deep pockets, they are forced by the model of decision making under consideration to spend vast sums of time and money to gather evidence of great harm only to have the evidence discounted later because, the argument goes, the likelihood of harm's occurring is so small as to be negligible.

I do not think that this model of decision making is the reasonable one to use in regard to many environmental matters—even though it seems quite reasonable and even though I thus seem unreasonable to object to it. It is one feature of our current natural social artifact regarding environmental matters and waste in particular, and it is no accident that it reinforces the status quo. It is symptomatic of a decidedly cautious cast of mind, one concerned *not* to err, one concerned not to invest heavily in what, without full information, cannot be fully determined by the facts, but must, on this view, therefore be determined by wish and by hope. I argue that what may seem necessary and reasonable in fact presupposes a cast of mind that is not rational, given the situation we often find ourselves in. The alternative is not to drive our decisions by wish and hope but by a different, but equally cautious and reasonable, cast of mind. I hope to convince the reader that in adopting a *properly* cautious cast of mind, he or she will not be giving up on reason, but using it properly. With a proper use, our current natural social artifact will be altered. It will be pulled into a new shape by a new mode of decision-procedure.

The point I am making is complicated because I am claiming both (1) that the current natural social artifact we have embodies the particular decision-procedure I think needs to be displaced and (2) that I want to concede, and do not know how anyone could not concede, that the particular decision-procedure to be displaced has its proper place.

The thrust of the second point is that we have taken a mode of proceeding that is quite reasonable in certain contexts and carried it, unthinkingly, into a context that can be made to fit it (which, indeed, it *will* fit as the next context will be conceived of in a way to guarantee a fit). The aim then is to show that the new context is more appropriately understood in a way that calls for a different decision-procedure, the one I articulate in this chapter.

The thrust of the first point is that the particular mode of proceeding to be assessed is part of a natural social artifact, the way in which we currently manage our waste and consider environmental issues. If we change the mode of decision-procedure, that will not be an isolated event but will ultimately cause a change in our natural social artifact, displacing it with another in tune with, or at least more attuned to, the new decision-procedure.

That changing our mode of decision making will change our natural social artifact of waste management, in particular, is a causal claim, and so subject to error. But the claim is motivated by the presumption that our current mode of decision making is so central to our current natural social artifact, permeates it through and through, working to maintain it against change, that changing the way we make decisions cannot help but have reverberations throughout that artifact.

I begin by considering what happens if we limit ourselves to interests and argue that, under that restriction, the decision-procedure I recommend is the rational and moral one to adopt. I make the restriction because I want to meet on the *weakest* ground those who might object to this new decision-procedure and show that even then, it is the preferred option as it answers our interests better. I then argue that if we consider the weight of rights in the calculation, there is no longer any contest. I must begin, however, by explaining how it is that interests enter our calculations when we assess our natural social artifacts.

Assessing Natural Social Artifacts

We can assess natural social artifacts the way we assess an action. If I act or fail to act—sharpen a pencil, plant daffodil bulbs, move a bush away from a walnut tree, leave the hornets' nest alone—I produce or intend to produce an effect—a sharpened pencil, daffodils, a healthy bush, a rash of hornets about. To assess a natural social artifact we need to track the changes that have occurred because of what we have done, evaluate them, and then evaluate the acts and omissions that led to them, all with a view to some intended effect or effects.

So, for instance, if we are to assess our practice of land use, to use an example from Chapter 2, we have to look at what people do to the brush and trees that block their views of the rivers and streams they live beside, determine the effects of their actions, and then evaluate those effects, asking, for instance, whether we want shallower and warmer streams and rivers. If we want those effects, we can then ask whether the homeowners' approach was the best way to proceed or whether some alternative would have been better; if we do not want those effects, we can ask what needs to be changed to prevent their occurrence. And about any changes we might propose, we need to ask after their effects, reiterating the series of questions as long as is necessary.

In short, we need to act the part of a stranger from another world and ask the following questions: (1) What changes have occurred because of what we did or did not do? (2) Do the changes make things better or worse,[1] or leave matters the same? (3) Whether things were better or worse, were the actions or omissions that produced those changes sufficient, or must we look for alternative causal factors that are necessary to produce the effect or effects? (4) Whatever the changes we might suggest, what changes will occur because of them, will they make things better or worse, and so on?[2]

These are standard sorts of questions to ask about any action or practice, but they make matters appear easier than they are. We know from Chapter 3 that the first and third steps, which seem be straightforward and empirical, are anything but that given the state of our knowledge, or ignorance. Such complications increase as we examine the future course of a natural social artifact rather than fix on a past configuration, for we must then make predictions and projections, oftentimes with little data. We can see as well that the second question requires an evaluation and thus some standards on the basis of which to evaluate. What end or ends do we wish to achieve? And how best can we achieve them? We obviously cannot answer such questions without evaluative standards.

For many things we do, the standards of evaluation are relatively straightforward and clear. In competitive tennis, the standards are determined by the end of winning the game: A serve is *not good* if it is easy to return, *is good*, or at least not bad, if it bounces high and to the backhand, and so on. Planting daffodil bulbs has gone well if the bulbs grow, bloom, and prosper over the years: The end is not to feed the squirrels or to bury the bulbs but to see daffodils. When we fail to achieve such clear ends, we look at alternatives to the action to determine if there is a *better* way to produce the effect or effects intended.

But, unfortunately, for many natural social artifacts, the ends are not at all clear. They may themselves have an indeterminate vagueness that prevents us from specifying exactly what we want to achieve and from being clear-headed about how to achieve it. And even when we can specify what we want, we may not specify it clearly enough to know where we have disagreement. We may agree that we want clean, unpolluted water, yet have very different conceptions of what that means, one thinking of the clarity of water in high mountain streams, another of the purity achieved through clorination. Each would find the other's conception unacceptable if each were clear-headed enough about what they vaguely had in mind to specify its details.[3] So we may think we agree on an end when we do not.

We may, indeed, have a multitude of ends we would like to achieve, and achieving some may make it more difficult, if not impossible, to achieve others.[4] It may be disputable what the end of a particular natural social artifact is, in other words, or disputable what the end ought to be. The current disputes over the virtues of incineration versus those of recycling are an object lesson in the contestable nature of ends. If we want to use waste to make energy, then the economics of running incinerators may make it less desirable to reduce the amount of waste produced.[5]

We cannot, in short, easily settle what the ends of waste management ought to be, what it *ought* to be designed to do. So this series of four steps gives only a rather specious clarity to our project. On top of that, it is difficult to get outside a natural social artifact and thus get away from its assumed standards. If the point of courtesy is to show respect for social superiors, those within the natural social artifact may well find incomprehensible not only any attempts at courtesy for other reasons but attempts to replace courtesy with other forms of social interaction. A revolution that displaces hierarchical social and political relations with democratic and egalitarian ones necessitates a new social artifact in which respect because of social position is displaced by "new social bonds of love, respect, and consent" based on the equality of citizenship.[6] Those within the older social artifact are ill-placed to assess their own practice by an appeal to such new standards as, for example, the essential equality of citizens.

We can get some purchase on assessing our own practice of waste management by focusing on the *interests* of persons. An interest is a stake we have in something so that, for instance, I have an interest in my good health because I stand to gain or lose depending on my state of health.[7] My interest in my health is set back if a municipal landfill pollutes the ground water from which the muncipality draws my drinking water: I am worse off because my health is worse off.

It is not that interests are somehow neutral—the hard data, as it were, of natural social artifacts, uncontaminated by them. Even how our interests are conceived is determined by the relevant natural social artifact.[8] But speaking of interests can begin to give us a purchase on what is common to a set of artifacts. It allows us to probe at the natural needs that are causally relevant to their production. We can speak of *common* interests in communication, for instance, or in health, and begin to assess natural social artifacts on how well they satisfy those interests.

We must, of course, distinguish interests from passing fancies and wants. What I want, transiently or even deeply, may not be in my interest. And even if what I desire deeply would cause deep satisfaction were it realized, I have no interest in it if I do not have a stake in it. Its realization may make me feel better, but the failure to achieve the end is no harm.[9] The test to determine whether one has an interest and its extent is to ask whether one's interests will be set back, and to what extent, if the end is not realized—whether one will be *harmed*, and how much. For some odd reason, I may deeply desire that the winning lottery ticket match my birth date and some other significant numbers, but if I have purchased no ticket and have no other stake in the outcome, my interests are not harmed if the winning ticket fails to match my desired numbers. Getting the number right would satisfy a want, not an interest.

We must also distinguish interests from one another, for the stakes can be higher or lower. I have a greater stake in being healthy than in having the neighbor's dog be quiet so I can work in peace, even though I hardly think about the former and often think about the latter. So, for another thing, having an interest in something in this sense does not imply thinking about it or even being interested in it. I may thus have interests in matters I know nothing about, such as a bill before Congress that would effectively lower air pollution even though I am ignorant of the bill's introduction. That is an interest others may share, and thus many of one's interests are held in common with others.

Each of us has an interest in safe drinking water, for instance, but before municipalities created reservoirs to provide such water, one person's realizing that interest might well have harmed the parallel interests of others—by tapping all the water of a spring, for example, or by overburdening the limited ground water.[10] To prevent such problems, and to guarantee pure water to all within a muncipality, we have a common interest in a municipal system that would satisfy our individual interests in safe drinking water without causing such conflicts of interest to occur. Water rationing in times of drought is meant, among other things, to preclude such conflicts.

Of course, what would satisfy such a common interest is highly likely to set back the interests of some individuals. Providing safe water for all may mean that some end up with less pure water, for example, than they had to begin with.[11] But we strive for the best of solutions, the ideal where satisfying a common interest satisfies the interests of each person and sets back the interests of none, despite knowing that achieving such an ideal solution is a rarity, even in situations where we all share a common interest. Many of our interests regarding waste fall into this class of common interests—our interests in clean air, in having the environment free of pollution and health hazards, and so on.

If we look at our object of concern, the natural social artifact that is our practice of waste management, we can ask the following questions: (1) What has occurred regarding our interests because of what we have done or did not do? (2) Has what has occurred made things better or worse for our interests—set them back, advanced them, or left matters the same? (3) Whether things are better or worse for our interests, were the actions or omissions that produced the changes sufficient, or must we look for alternative causal factors that are necessary to produce the effect or effects? (4) Whatever the changes we might suggest, what changes will occur because of them, will they set back interests (and if so, which ones), advance interests (and if so, which ones), and so on?

Taking a Risk

One interest affected by our current natural social artifact is our common interest in a healthy environment. Just as there is little doubt that one of the most fundamental advances in health care occurred at the end of the nineteenth century when water and sewage systems were installed, so there is little doubt that a major setback to that advance has occurred within the last twenty years or so with the pollution of our water supplies—by infiltration of pollutants into ground water[12] and by air pollution that so harms the rain that it cannot be captured and drunk without purification.[13] In addition, we have to concern ourselves with agents associated with a higher incidence of cancer and other diseases[14] being introduced into the environment—from dioxin being spread on roads to keep down the dust[15] to weed killers for suburban lawns and public parks and roadsides getting into the air and leaching into the water[16] to lead being used in ceramic dishes[17] to medical waste being dumped at sea and washing up on our shores to various pesticides getting into our food supplies.[18]

What we know is frightening, but, worse, we have little knowledge of the full impact of our normal methods of disposal on our health. As we saw in Chapter 3, we do not fully know what changes have occurred—how badly, for instance, our ground water is contaminated, if at all, or how long it will take to cleanse itself, if it can. And we can easily find object lessons in our inability to make firm predictions. One example may be the recent discovery that many of the estimated 47,500 55-gallon barrels filled with plutonium, mercury, and cesium dumped in what is now the Gulf of the Farallones National Marine Sanctuary are corroded and have burst under water pressure.[19] What we do know is that we may put our health at presumptively great risk by our current methods of disposal. Because we all are presumed to have an interest in our health, one of our interests is thus presumptively harmed by our current natural social artifact.

I say "presumptively" because I want to emphasize that our ignorance extends as far as the possible harm to us from current practices and because that ignorance creates space for an apparently more powerful variant of the objection considered in Chapter 3: When we do not know what harm, if any, is being done to our interests, how can we justify vast expenditures to change our practices? No one may actually be at greater risk, and, the argument continues, in any event, to show that someone is at risk is not to show actual harm. Even if the risk to our health is greater with our current practices of disposal, for instance, it does not follow that anyone's health is actually worse off. The risk of a disaster at a nuclear plant is obviously greater than it was before such plants were built, but it does not follow from this that our interest in having a radiation-free environment is actually being harmed. Without a showing of actual harm to our interests, the objection concludes, it is unreasonable to remove the risk at great cost to other interests, such as cheaper and cleaner energy from nuclear reactors.

There is a power to this objection, for it depends on what appear to be—but are not, I shall argue—two obvious truths. First, to be at risk does not imply being harmed; and second, acting without all relevant information is, to that extent, unreasonable. The two apparent truths combine to produce what appears to be a decisive objection. How can we justify spending vast sums to change our practices when we know, at best, only that we are at some risk? If our ignorance is a crucial variable, the objection can be put even more powerfully. How can we justify harm to our interests in sustaining our current natural social artifact, with all its entrenched capital expenditures and habits of use, when we are ignorant of the potentially harmful effects of the available alternatives? Because being at any risk does not imply being harmed, how can we justify any action at all to mitigate the risk if such an action itself harms our interests?

Put that way, the objection begins to sound odd. "Surely," we want to ask, "it is reasonable to reduce risk when we can?"[20] If not, then because any significant action regarding waste would require a change in our practices and we have an interest in maintaining those just because they are our practices, we should not act at all until we have evidence of actual harm.

But when we are at risk, our whole way of moving through the world changes, for we become nervous and cautious. It is one of the *real* harms of an increased level of violence on city streets that inhabitants perceive the risk to be greater and that they shun the streets when they can and travel carefully when they must travel. It is obviously arguable that being put at risk can harm one's interests, and, as I have argued, we can act reasonably without all relevant information.

Yet, I do not want to expound the second argument again or pursue the first. The objection to which they are responses reflects a cast of mind, a mode of procedure, that we need to displace with an alternative. It is a mode of thinking that emphasizes the open-ended nature of scientific problems, the appropriateness of asking if we are sure we have all relevant information. It is a mode of thinking that emphasizes caution, as I do, but caution, as it turns out, in the face of vested interests, noting how often we have moved in error and how much harm can be done to existing interests if we act precipitously. Showing up the weaknesses of this cast of mind helps, but the cast of mind has strengths, especially in the face of ignorance and of harm to current practices if we act, and its problems will not in themselves restrain its continued use in the face of mounting environmental problems. We must show that there is an alternative rational and moral way of coming to decisions.

The bottom line to assess any difference between that cast of mind and an alternative is that we would make different decisions. If we had followed the method I propose, we should not have had the *Exxon-Valdez* disaster: Even the small possibility of such a disaster would have weighed too heavily in the calculus to justify putting an oil tanker terminal in such a place. But to make this clear, I need to examine the idea of acting cautiously, using two examples—polybrominated biphenyl (PBB) and chlorofluorocarbons (CFCs). The aim is to provide relatively clear cases where, even with great ignorance, it is reasonable to act and reasonable not to act. If these two cases are clear enough, they can become benchmarks by which we can measure other less clear cases.

Clear Cases

When PBB was discovered in the environment in Michigan, there were, as we saw, immediate outcries to remove it, by condemning the livestock that had ingested it, for instance. The Michigan Department of Agriculture condemned milk that had been adulterated with PBB, but most other food was left untouched.[21] The various governmental agencies involved delayed acting about some matters for some time, changing its position, as we saw, of what amount of contamination by PBB was enough to justify banning foodstuffs, refusing to prohibit the export of cattle, for instance, that had some level of PBB contamination, and so on. By their apparent foot-dragging, the governmental agencies involved left themselves open to the charge that they delayed action just because they knew that as the PBB spread and so was dissipated into increasingly minute quantities, it would become more and more difficult to test for and the problem would disappear. I have argued that the concern not to harm current interests without a showing of real harm exerted a gravitational pull on the relevant governmental agencies, slowing their response significantly. But the slow response, however it was motivated, was arguably not an unreasonable one in regard to at least *some* of the effects of PBB—such as its dispersal in various farmyards even though that entailed a subsequent risk to ground water and to our food supplies because, for instance, chickens ate it.

The harm was caused by a single load of PBB being introduced into animal feed. There was little danger of that sort of accident occurring again, and steps were quickly taken, by marking bags more clearly, among other things, to ensure that it did not happen again. So the various governmental agencies were not faced with an ongoing situation that would worsen if they failed to act. They faced a situation that would probably get better as time went on, as the PBB spread farther and farther and its concentrations were diluted as it dispersed more and more. So time was arguably an asset in regard to some of the effects.

In addition, not everyone's interests were affected. We did not have a global crisis and would not have had one even beyond portions of the state of Michigan if the state had acted quickly to isolate the infected livestock and prevent its movement onto the market.

Just as important, it was unclear what could be done to remove PBB from the environment. It was spread widely and thinly in silos and feed lots, in such minute concentrations that it tested the limits of detection equipment. And because no one had ever tried to clean it up from the environment in such a form before, no one had much of an idea of how we could get rid of it—what the insides of silos could be cleaned with to scrub off the residue, for instance, whether workers would disperse it into the air as they cleaned, whether they would breathe it in or absorb it through their skins and thus whether working with it would be dangerous for them, and so on. State authorities thus faced an expensive and labor-intensive enterprise, one that was arguably more risky to those doing the cleaning than to anyone else, with little idea of whether such an enterprise could be a success.

I do not want to get involved in the details of how the PBB problem was handled, with whether the agencies involved delayed action out of a reasoned response to the situation or out of normal or abnormal bureaucratic tardiness, for instance. Nor do I want to get involved in assessing all parts of the state's response, whether, for instance, the state took due care to protect consumers in Canada and other states in failing to prohibit exports of suspected meat and poultry products and took due care to protect consumers in Michigan. I want rather to emphasize that the response to at least parts of the crisis could be *less* cautious about possible harm to citizens and the environment in part because as time went on the agent causing the harm was being cleansed out of the environment or at least diluted to the point of negligible effects, and because anything that could be done would be very expensive, with unclear chances of success. It was not our ignorance of the harmful effects of PBB that justified state inaction in regard to *some* of the effects of PBB, I hasten to add, but both our knowledge that even if it were known to be harmful the risk of harm was diminishing every day and our ignorance about quite how to cleanse the environment.[22]

The crisis with CFCs, however, is a different story.[23] The ozone layer protects us, among other things, from the harmful effects of the sun's ultraviolet rays, and the discovery in the early 1970s of a diminution in the amount of ozone over the Antarctic was thus a cause for concern. The working hypothesis was that the decrease was caused by a chemical reaction between ozone and CFCs, but when the issue first presented itself, three different sources of uncertainty plagued initial attempts at actions to decrease the amounts of CFCs being released into the atmosphere.

One source concerned the effects of CFCs themselves. How exactly were they responsible for ozone depletion—by themselves or in concert with other, perhaps more significant, causal agents, and in a way that would reach equilibrium at no or little loss of ozone, or in a manner that might accelerate after some crucial and unknown trip point? A second concerned the scope of the supposed effects. How much depletion can be expected given the amount of CFCs in the atmosphere at present, and how much given the projected increase? The third uncertainty concerned the effects of ozone depletion itself. What effects can be expected for us and for other life forms, from plants and primitive animals in the food chain to higher forms of life that may be affected not just by the ozone depletion directly but by its accumulated effects?

With the help of entrenched interests and an entrenched stance toward how to make proper decisions regarding such matters,[24] these three areas of uncertainty conspired to prevent any significant action for some time. We may divide the use of CFCs into nonessential and hard-to-replace, the former marked by its use, for instance, as an aerosol propellant, the latter by its use in certain plastic foam products, as a cleansing agent in manufacturing computer microchips, and as a refrigerant. The ease of replacing its use as a propellant is matched by the difficulties of finding replacements in its other uses. In either case, nothing could be done about the CFCs currently in the atmosphere, and because they had not produced a significant decrease in ozone, caution in acting even regarding nonessential uses apparently did not seem unreasonable. If we may project a reasonable narrative back on to a past that seemed at times to defy reason, a consensus was reached that a measurable depletion of ½ to 2 percent—judgments varied—would be cause for severe concern.

The sudden and unexpected discovery of a decrease of almost 40 percent in the ozone over the Antarctica between 1977 and 1985 made such depletion projections irrelevant and trumped the sources of uncertainty: In the face of such a loss, even with uncertainty, inaction was unconscionable.[25]

It is important for our purposes to slice this process sharply at two different points, in 1975, say, and the present. *Now* there is no difficulty justifying the banning of CFCs in aerosol sprays and in making every effort to minimize their uses in other applications, even in the face of the same uncertainties we faced in 1975. We still do not know what effects we may anticipate from the depletion of ozone: Will the earth's life forms be under intense pressure, and if so, how will they respond?[26] We still may not understand completely the mechanism of depletion, or even why the depletion occurred over the Antarctica,[27] and we still do not know where we are in the depletion story, how much more can be expected or whether levels may revert to normal.[28]

Yet, if we project ourselves back to 1975, say, we may ask, given the *same* sort of uncertainties, but without the later evidence of the 40-percent decrease over Antarctica, "What is it reasonable to do?" To answer that question properly, we must bring in the other relevant variables that mark the ozone story.

For one thing, the difficulty is of a continuing nature. We not only had an amount of CFCs in the atmosphere that could not be recovered, but also their release into the atmosphere occurred almost every time someone used a can of hair spray, shaving cream, deodorant, or insecticide, for it was the propellant of choice among manufacturers of these and similar products.

A second variable is that CFCs are beneficial. No one thought that PBB served any beneficial function when ingested, and so no one argued that harm to our health would result from its removal from our food chain. But although CFCs need not be used as propellants in aerosol sprays, their use in a liquefied form by the electronics industry to clean computer chips and other micro-electronic parts and their use as insulation and in refrigeration products are valuable. Chlorofluorocarbons "are stable, nontoxic and cannot catch fire,"[29] and so when they displaced other such refrigerants as ammonia and sulfur dioxide, they were a real benefit in increasing the safety of refrigerators, for instance, which sometimes exploded and sometimes leaked poisonous fumes. In addition, CFCs have such good insulating characteristics that their replacement in the United States alone is expected to "increase electricity consumption . . . by up to 94 billion kilowatt hours a year, about 3 percent electricity more than used now" in the United States.[30] So banning CFCs carries a real loss if they are not replaced with other products equally efficient and useful.[31]

A third variable was that the problem was international, affecting the entire earth and all the earth's life forms. We could in 1975 only speculate on the effects of an increase in ultraviolet radiation, but a worldwide increase in the incidence of skin cancer is not an unreasonable prediction.

At any rate, the damage likely to be done could not be isolated, as in the PBB crisis in Michigan, but would affect everyone in the world—whether they had the benefits of CFCs or not.[32] The assumption is that a decrease in ozone over the Antarctic signals a general decrease in ozone over all the world.

The scope of the problem and the clear benefits of CFCs complicate a solution. The United States produces a third of the two billion pounds of CFCs used each year, but a significant reduction in the amount produced in the United States would significantly reduce the risk from CFCs only if other nations did not take up the slack. Because CFCs are so beneficial, reducing production in the United States without providing an alternative will drive up demand, making it profitable for new companies in other countries to enter the market. For this reason, a single nation is almost as politically helpless facing this problem as a municipality trying to regulate the bottles of international bottling companies. Again, the situation is unlike that posed by PBB in Michigan. Because there was no demand for PBB as a foodstuff, what was needed was to clean it up, not provide a substitute for it, and the scope of the problem, though large, did not need the sort of international cooperation and all that is entailed regarding negotiation and enforcement required by the problem with the ozone layer and CFCs. So one variable that makes a great difference in the CFC problem is the form any solution must take.

But the crucial variable, I would suggest, is the odd combination of certainty and uncertainty itself—the certainty that we face the *possibility* of catastrophic harm to us and the rest of the ecosystem and the uncertainty about how CFCs work, including the uncertainty—as the discovery over the Antarctica made dramatically clear—about how quickly they can work.[33]

In such a situation, with such a set of features, it begins to approach the unreasoning insistence of a willful child not to act because no *actual* harm has been shown.[34] As an EPA official said about the situation early on, "If we wait until 1990 to make a decision, it could be too late." The problem is that "from five to ten years might pass before sufficient data could be acquired to conclusively prove the theory of ozone depletion by chlorofluorocarbons," and "all the chlorofluorocarbons produced in that period would make their way into the stratosphere."[35] Waiting for proof that actual harm has occurred will guarantee that actual harm has occurred, and because increased use would curve upward with any inaction on our part to slow the use, much more harm would occur than would have otherwise occurred.[36]

Delaying to act at least in regard to some of CFCs' uses was especially inappropriate because in 1975, as today, an alternative or set of alternatives existed that could serve the same functions the culprit served—though not without harm for some of its uses and not without hardship to manufacturers and consumers. Different propellants for aerosal sprays are relatively easy to produce, for instance, and there is no compelling need for CFCs as propellants.[37] It cost manufacturers little to change the way in which sprays are propelled, and the effect was a significant reduction of any new CFCs entering the atmosphere.[38]

Replacing CFCs in their other uses is more complicated. One virtue of CFCs is their inherent chemical stability: They are bonded together with the strongest chemical bonds possible.[39] So CFCs do not readily deteriorate and do not readily degrade into anything harmful. They last longer than the refrigeration equipment and insulation they are used in. But this virtue is a fault. Because they outlive the products in which they are used, they are released into the air and are stable enough to rise to the stratosphere where intense ultraviolet radiation begins to do them in, initiating the process that leads to ozone depletion. Any alternative to CFCs may lack the virtues conferred on it by its stability. More products may lose their insulating properties more quickly, for instance; thicker forms of substitute insulation may be needed; and the currently available substitutes—HCFCs, which are hydrogen compounds—deteriorate more quickly than CFCs but are not wholly benign. So consumers may find their appliances bulkier, larger on the outside or smaller on the inside, and may find that the insulation sometimes fails before the appliance. And although there will be a net gain in stopping the loss of ozone, the world will not be pristine. Manufacturers will have to phase out equipment used to make CFCs, at a cost, and invest in new equipment to make alternatives, at a cost.[40] So, as usual with such matters, no solution provides a free lunch, but alternatives to CFCs exist, and existed in 1975, that cause significantly less harm than it was feared CFCs could cause.

These features conspire to provide a solution. Had we known in 1975 that our fears about the extent of ozone depletion would be so far short of the mark, we would surely have acted then.[41] But our ignorance in 1975 does not change the situation we faced then: Given the variables, it was more reasonable to act than not to act.[42]

The Decision-Procedure

The variables in these benchmark cases work together in complex ways: the amount and kind of harm that could occur, the risk that such harm would occur, the extent of our knowledge of the causes of the harm, the persistent or transitory nature of the cause, the necessity of the cause and the ease or difficulty of finding alternatives to the cause, the state of our knowledge of the consequences of alternatives, and so on. The extent and degree of our caution, and thus the extent and degree of our action or inaction, will vary from case to case. The CFC case today is ideal for action (in which failure to act begins to approach utter unreasonableness), and in 1975 it certainly approached the ideal for action. The PBB case tends closer to the ideal for inaction in at least some of its aspects.

About neither do we have much idea, if any at all, of how the causal agents work or even whether they would cause harm. We may not understand fully the mechanism by which CFCs, or some of their elements, break down ozone or affect something else that may break down ozone. We do not understand fully the mechanism by which PBB is taken up by organisms and deposited in the fatty tissues. We do not know how much of a health hazard PBB is to humans, and we do not know for sure that the breakdown of ozone is caused by, or solely by, CFCs, whether it will continue even if they are reduced or will reach some natural balance, what the full effects of its loss will be, and so on. Yet none of that matters in deciding what it is reasonable to do.

For in our determining what is reasonable the various relevant variables take on more or less importance, more or less weight, as each varies. If the harm is very great, then, in determining whether it is reasonable to act to prevent it, we do not demand for action that great a likelihood of its occurring: Even a small chance of a very great harm is considered sufficient for concern. If a remedy to remove the harm is easy and known to be without harmful side effects, we are less concerned about our ignorance of the real causes at work and more moved to act the more or the less we know about the harmful or benign effects of the remedy. If we are ignorant of the actual mechanism that produces a harm but have some evidence linking it with a supposed cause, then it is more reasonable to act to remove the cause if the harm is irreversible. If the harm is known to be very great, but only to a relatively trivial interest of ours, then we think it less reasonable to act. If what we need to do to change our practices causes moderate harm to our current interests, then we demand a showing of a greater risk than we might otherwise, or a greater harm, and so on.

There is no formula here, no measured and quantified way of taking into account all these variables and mapping out what happens when the possible harm doubles, our knowledge is decreased by a third, or the cost of change quadruples. As Aristotle says, we can expect no more precision than the subject matter admits. One consequence is that in some cases we will not know what it is reasonable to do. The variables will conspire to leave us half of one mind and half of another. In such a case inaction until we obtain more information may be the most reasonable response. In other cases the harm may be of such a magnitude, though only possible, and the alternatives so easy to institute, that we would think it approaching utter unreasonableness to counter with the claim that no harm has actually been shown. In fact, the usual case, I would argue, is one in which we will be able to determine clearly enough what ought to be done—especially if we did not have to concern ourselves with the interests of special interests. For although we may lack information about some variables, other variables conspire to provide a basis for action or inaction.

For example, even if CFCs are not wholly at fault, if the risk is not as high as some think, and if we do not know whether the breakdown of the ozone layer will now slow down or whether we have gone past the edge of a slide in which nothing we can do will stop its further deterioration, making any present action useless, still, we do not need to use CFC gases as we have used them. We can act to rid the environment of additions to what presently occurs, or at least significantly lower their occurrence, without undue harm to any of our other interests.

It is not that no other interests are harmed by our actions. Manufacturers have equipment set up to handle CFCs and to manufacture them for use as propellants, for instance, and the replacement or refitting of this equipment has a cost. In addition, what is presently the most reasonable alternative to CFCs, HCFCs, also contain chlorine, the chemical responsible for the depletion of ozone, and although their introduction is a big gain because, as F. Anthony Vogelsburg, environmental manager for DuPont's Freon Products Division, says, "They have one-tenth the lifetime and one-tenth to one-twentieth the ozone-depletion potential" of CFCs, they are not benign,[43] and a provision to ban HCFCs after 2010 was expected to be introduced when the Senate acted to renew the Clean Air Act.[44] But, as Vogelsburg notes, DuPont would have to invest $1 billion to replace its share of the market for CFCs, and worldwide $4 to $6 billion would have to be spent. It takes at least twenty years to recover such heavy start-up costs. So banning HCFCs after 2010 means that no new plants will be built after 1990 and thus that no new plants will be built, unless new incentives are introduced into the calculus.

If Vogelsburg's figures are correct, this may be a situation where the losses incurred by these companies should be shared in some way, and one good reason for subsidizing them is that the nature of the harm has a heavy weight in a proper decision-procedure for determining what to do. Another good reason for subsidizing those who will lose out is that they are not at fault. The introduction of CFCs was not the result of some conspiracy to degrade the environment and cause anyone harm. To penalize manufacturers for developing useful products using chemicals and other resources that no one could have known were harmful at the time of their introduction would add a heavy cost to innovation.

On the other hand, we should not be too quick to subsidize those who may lose money through such regulation as that which prohibits the use of such chemicals as CFCs. Forcing old ways to be changed can sometimes produce beneficial effects, and although these benefits are about as unpredictable as the total effects of ozone depletion, they must be weighed into any calculus that considers compensating those whose entrenched interests will be set back.[45]

Those interests will be harmed, but not nearly as much as the interests of everyone else. For us the possible harm is very great indeed, and it is harm to the interests of all mankind, a global and not just localized harm. Given what evidence we do have that CFCs are responsible for ozone depletion, a compelling reason exists for overriding interests that may be harmed by changing manufacturing requirements. The questions to ask are whether CFCs are necessary, and if they are not, whether there are alternatives, and if so, whether the alternatives are better—and not just better, but better enough to justify disruption of our current practices. That is, in this case, are there alternative ways we could gain the goods achieved by the use of CFCs—the convenience of packaging liquids in cans, for instance, and tidy and efficient refrigeration units—and do those alternatives cause enough less harm to our interests to justify disrupting our current practices and thus harming our current interests?[46] This question approaches the rhetorical in the case of CFCs.

On the other hand, scrubbing out the residue of PBB from silos and storage facilities is an enormously expensive and complicated undertaking, if it is even feasible. But here there is little risk of additional PBB being put in the environment in the way it was, and what is there will eventually dilute to the point where one can no longer measure its occurrence.

So when I said in Chapter 3 that we should act cautiously, I meant that the degree of caution should be tailored to the situation and thus to the other relevant variables. In the case of CFCs, the risk to health and to the environment is enormous, but we can act, even without all relevant knowledge. In the PBB case, not acting, at least in regard to some matters, was arguably the more reasonable response. Both cases urge caution: Do not do something that may cause more harm. But the appropriate responses are different in these cases because of the ways in which the relevant variables are arranged. The source of harm was transitory in the one case and is persistent in the other; the nature of the harm is universal and may be devastating in the one case and was contained in the other; and so on.

So What Is New?

The mode of thought being recommended must face, in the public arena, demands that seem quite reasonable—a demand for a showing of real harm, for instance, to justify action that would itself cause real harm to existing interests, or a demand for more information in a situation both where we do not want to cause unnecessary harm to existing interests and where we should not precipitously create a new set of interests. Against those demands, it may be unclear as yet how that new cast of mind is to work. How would the structure of a public policy discussion dominated by this recommended mode of thought differ from one dominated by such apparently reasonable demands?

The *Exxon-Valdez* case is a good one to examine. The actual course of the public discussion about whether to allow drilling on the North Slope of Alaska and, if so, where to run the pipeline that drilling would require was complicated and convoluted, and the various interest groups and competing interests are difficult to distinguish, let alone track through the tortured meanderings of claims and counterclaims. But we need not provide a detailed anatomical description of the course of discussion to discern its logical structure, for the claims and counterclaims all took place within a clearly confined set of permissible moves.

Consider, for instance, the discussion about whether to place such a facility as a tanker terminal in such a location as Prince William Sound. On the one hand, it was argued that any harm that might occur because of a spill would be minimal, retained by booms on the spot, by the peculiar protected configuration of the harbor, by the absence of winds, and what have you. On the other hand, it was argued that even if one envisaged a spill that could be catastrophic, an oil tanker running aground for instance, the chances of such a spill occurring were virtually nil. It was thus reported that "environmental studies and state reports dating almost to January 1968, when huge oil reserves were discovered on Alaska's North Slope, predicted that a major spill was so unlikely that planning for one would be unnecessary."[47]

We can document the pathology of such discussions by noting the same moves being made, over and over, in other public policy discussions. First it is claimed that a failure would not be harmful. Then it is claimed that even if a failure were catastrophic, it is extremely unlikely to occur. We can get caught between these moves. In response to the first, we might try to show that, indeed, great harm could result from the proposal, whatever it is. All a municipality's groundwater could be contaminated, or low-level radioactive waste could increase the incidence of cancer and other diseases as its containers aged and leaked, and so on.

But then, after a laborious process of showing such harm, we are met with the apparently reasonable, "But it is so unlikely!" and that ends the matter. The discussion is pinched between the reasonable request, "Would a failure be harmful?" and the reasonable response, "But is it likely?"

Such a move is often the public policy equivalent of bait-and-switch tactics. We may feel baited into a long and expensive investigation into how great the potential harm might be, knowing well how difficult it is to *prove* that real harm will occur in arenas with so many variables and so many projections of probabilities and knowing that the very form of the investigation will leave many loose projections for opponents to climb on. On the one hand, then, we commit oneselves to an expensive and time-consuming investigatory process that will provide opponents with footholds for objections and so real bases for continually playing the epistemological card. But after a showing of that great harm, the ground of the debate is switched so that a showing of great harm is unnecessary: If a failure is unlikely and that is a sufficient justification for building an oil tanker terminal, for instance, what does it matter if a failure were catastrophic or minor? Yet, the tactics are dictated by a cast of mind that seems reasonable. We should not cause real harm now, to existing interests, unless we can show that real harm will occur if we do not change our practices.[48] One must meet two demands: show significant harm, and show that such harm is not just possible but also likely.

The mode of thought I am recommending cuts the connections between these two demands just as it cuts off as inappropriate, in some circumstances, the otherwise reasonable demand for more information. In so doing, it should change the structure of public policy discussions, for if we show that the possible harm is catastrophic, as in the case of CFCs, that in itself has great weight in the calculation of what we ought to do. This recommended cast of mind requires that we treat worst-case scenarios as *real* possibilities when the worst case causes great harm. Whether that has sufficient weight to determine our public policy will depend on the other relevant variables and on its overcoming the inertia of past practices protected, as they are, by real political barriers and by real interests that would be harmed by change. Yet, the form of dependence does not require that a harm be *both* catastrophic and likely. If the harm is sufficiently great, concerns about its low likelihood diminish.

Thus the siting of an oil tanker terminal in an area as pristine, as productive, and as land-locked as Prince William Sound where the *Exxon-Valdez* ran aground raises such questions about the magnitude of harm in case of an accident that concern about how likely or unlikely such an accident may be ought not to weigh heavily in any decision about what to do. Given the catastrophic consequences of an accident, *should* it occur, we should be extremely reluctant to site an oil tanker terminal there.[49] This reluctance can only increase when we factor in our knowledge of how easy it is for mistakes to occur in any complex practice such as docking, unloading, and maneuvering a tanker and recognize how often mistakes have occurred when they were thought "so unlikely that planning for one would be unnecessary."

The proposed decision-procedure may seem so obvious that it is hardly worth noting: "How could anyone disagree," we may wonder, "with a procedure that takes very seriously the great harm that *could* result from an accident?" But, as I have tried to emphasize, the current mode of discussion ties together a calculation of possible harm with a projection of its probable occurrence, downgrading the importance of the magnitude of the harm if it is thought unlikely, and that tie can seem perfectly reasonable—especially when change itself would harm the interests encased in the current practice. That tie also has great practical effects.

Consider, as another example, the discussion about the greenhouse effect. This discussion has been dominated by two primary concerns: What sorts of harm are likely to be caused, if any, and what is the likelihood that there will be harm? About both these concerns there are questions, and if we enter into a discussion of the data regarding either, we enter into a black hole of claim and counterclaim, speculation and hypothesis, from which we cannot know whether light will emerge in time for us to know whether we should act to prevent harm. "Current forecasts of global warming," it is claimed by Richard S. Lindzen of MIT and Jerome Namias of the Scripps Institution of Oceanography, "are so inaccurate and fraught with uncertainty as to be useless to policymakers."[50]

What is known is that "first, carbon dioxide, the waste gas produced by burning coal, oil and wood, has been accumulating in the earth's atmosphere over the last century; and [that] second, the gas traps heat that is produced when the sun's energy is absorbed by the earth and then reradiated." What is in dispute is "how much the earth will heat up after the injection of a given amount of carbon dioxide" and what other effects the increase in carbon dioxide—and such other gases as methane—may have.[51]

On the one hand, it has been suggested that if there is an increase in carbon dioxide, that will allow plant life to flourish, and because plants take in carbon dioxide, eventually the amount of carbon dioxide on earth will reach a new balance. The increase, it is claimed, will produce hardier and more prolific plants. Sherwood B. Idso of the Agriculture Department's water conservation laboratory in Phoenix says, "I'm really sold on carbon dioxide; I think it's very beneficial."[52]

On the other hand, increased cancer, loss of land and island nations, and other harms are predicted. And the predictions of the likelihoods of such a package of harms and benefits occurring are as varied as the models of prediction used by scientists trying to make projections on very little data with very complicated sets of interactions. Fakhri A. Bazzaz, a plant ecologist at Harvard, says, "The consequences are unbelievably complicated and there are so many of them."[53] For instance, some experiments seem to show that some insects do not mature properly and have a higher rate of mortality when they eat plants grown with higher concentrations of carbon dioxide. The number of insects could thus be reduced, affecting higher forms of animals that live on insects and preventing pollination of plants. Even worse, it has been suggested that warming itself "produces more warming by stimulating the release of carbon dioxide from natural sources."[54] So whatever harms may result from warming will be magnified we do not know how many fold.

On neither question—"What is the harm?" and "How likely is it to occur?"—is there consensus in the scientific community. We normally use the scientific method as a way to winnow out subjective judgments and reassure ourselves that we have at least as much objectivity as can be obtained. But here the scientific method has not yet produced agreement. Indeed, it would be difficult to imagine that there could be agreement with such surprises as the heavy loss of ozone over the Antarctic being the rule and making for caution and with new data coming in regularly that cannot easily be understood and assimilated into current models and projections.[55] For example, it has been reported that OH, a hydroxyl radical consisting of an oxygen atom and a hydrogen atom, has been depleted sigificantly in the Northern Hemisphere, with the depletion being more than twice as extensive as had been thought. OH destroys methane in the atmosphere. So even though more methane is being produced, "less is being destroyed there naturally."[56] Why OH is being depleted is unknown.

It boggles the mind to try to determine what to do about the greenhouse effect on the basis of an assessment of the harm that will result compared with the benefits to be gained and of the probabilities of those various harms and benefits occurring. We cannot know that anyone, scientist or nonscientist, is in a position to know about these matters. So, the question is, what ought we to do given the failure so far of the scientific community to reach agreement?

One option is to do nothing and await the results of further research that will give us the knowledge we need to determine what to do. To set public policy *without* such knowledge, it is argued, is to impose "economically harmful policies to limit energy use by imposing special taxes or other restrictions."[57] It is tantamount, in short, to cause harm without showing that any real harm will be averted. Limiting energy use, it is claimed, will slow economic growth, among other things.

What lies behind this prescription for further research and public policy inaction is that powerful, but mistaken, conception of what is necessary if we are to act reasonably. It is the view that "it is neither reasonable nor prudent for major public policy decisions to be based on presumptions about issues in science, which, in the current state of knowledge, are still only hypotheses."[58]

We may take this as a straightforward epistemic condition for intelligent and prudent action: Know what causes which effects before acting to change any effects. But it is a condition backed by a concern not only not to do something unintelligent but also not to cause unnecessary harm. In arguing against banning CFCs, Dixie Lee Ray says, "Replacing *just* the refrigerated transport for food moving to market would cost more than $150 *billion*. Some estimate the cost of banning CFCs for refrigeration at $800 per person per year."[59]

Another option would be to do whatever can be done to minimize the emission of carbon dioxide on the belief that it is better to be safe than sorry. With the magnitude for potential harm, it may be claimed, we ought to act with as much dispatch as we can to minimize the possibility of a real disaster.

Neither of these options is based fully on the evidence available, I would argue. What drives both are competing presumptions about how to respond to possible risks of potentially great magnitude, for it is consistent with the evidence available both that we will have a great disaster if we do not curb the emissions of carbon dioxide and that we shall have no disaster at all. So the first option, in urging us to do nothing until we have more research, courts a disaster, and the second, in urging us to do everything we can to cut emissions, risks great harm to current interests for no good reason. The only way to make sense of these responses is to add some additional premises to the arguments each side gives—competing presumptions about how we should respond to potential harms.

There is a middle way between these two extremes of inaction on public policy and a full-press assault on carbon dioxide emissions. It requires that we reject the view that it is unreasonable to act without all relevant information and even more unreasonable when such action will harm existing interests and that we reject the view that it is unreasonable not to do everything one can to diminish a risk. It is a way that requires the cast of mind I have been arguing for here.

That we do not know that more heat being trapped in the atmosphere will cause more actual harm than good, that we do not know how much more heat will be trapped, for how long, and at what rates, and so on for all the other variables about which we are supposed to need knowledge—all that is irrelevant in deciding that we ought to act now, in the face of a harm that may be very great indeed, to do at least the *minimum* to decrease the amount of carbon dioxide and methane we are now producing (and prevent the loss of OH if we can).

That judgment has shaped a cast of mind that recommends acting to reduce such waste products as carbon dioxide and methane in the face of the magnitude of harm they may cause, and it is a judgment that needs to be played out in particular fields of activity (auto emission controls, reductions in the use of certain fuels, and so on) with a due concern for the harm to existing interests and the ease of alternatives to existing practices.[60] We could, for example, do much more to encourage passive solar power and a reduction in the use of gasoline, each of which would encourage less production of carbon dioxide. We need not get caught up in a never-ending discussion of possible and probable harms and benefits to make a decision about what to do, a decision dictated by a quite reasonable concern that the magnitude of harm that could occur is great indeed[61] and that the longer we wait to act, the higher the curve of potential problems and the longer the recovery period from the harm caused.[62]

We find again and again in public policy discussions the one-two punch of the form of reasoning that prevents such conclusions: "What could be the harm?" and "How likely is it to occur?" Severed from one another, as the separation of the questions itself suggests, this form of enquiry produces an oddly disjointed decision and a disjointed discussion of the problem. If the possible harm is high but the probability judged to be negligible, the process goes on, with the magnitude of harm itself somehow diminished to a trifle as a factor to be considered by the low probability of its occurrence or, rather, diminished by the *judgment* that its probability of occurrence is low *or* by the judgment that it is unknown and that action to change the status quo must await a better knowledge of likelihoods. We can find examples of such reasoning over and over in various public policy discussions, and, unfortunately, we must live with the residue of past decisions made when the judgment of the magnitude of harm was diminished by a judgment of the slight probability of its occurrence. The *Exxon-Valdez* disaster is just one of the most recent and notorious examples.[63]

In summary, the decision-procedure I propose forces us, in examining such cases as siting an oil depot on Prince William Sound, to concentrate on the magnitude of harm that would occur were we to suffer an accident we might admit is unlikely. Other cases might call for a concentration on other relevant features such as the scope of the harm that would occur, whether it is of a continuing nature or is self-contained, and so on. In any event, what is new in the procedure I recommend is that we cut the tie between the demand to show significant harm, on the one hand, and that such harm is not just possible but likely on the other. It is enough, as the *Exxon-Valdez* case and the case of CFCs and ozone are meant to illustrate, that there be the possibility of real harm to our interests, however unlikely it may be that the real harm will occur.

Presuming an Accident

In a county in Southwest Michigan, residents were concerned by a proposal to site a landfill on a hillside immediately adjacent to a river. The landfill could contaminate the river either through leakage into the ground water, if it flows toward the river, or through run-off and spillage or seepage. The highly spirited public debate turned on the likelihood of such contamination—whether it helped or hurt that the site was a former gravel pit and the hillside quite porous, whether new sorts of landfill liners would safely contain the various liquids and toxic substances such as mercury that might otherwise leach into the soil and thus, with some unknown degree of probability, into the underground water and, again with some unknown degree of probability, eventually into the river, and so on.

In such a case, we are hard put to assess the likelihood of various harms. The possibility of something going wrong occurs at too many junctures in a complicated chain of circumstances, with too many new variables being created by any accident, and with too little knowledge of the mechanisms involved, for us to make predictions of what is likely and what is not with any great degree of certainty. We are hard-pressed even to put a number to the degree of our assurance about any prediction.

In the usual case, however, we have enough background information to make a projection about the likelihood of harm occurring *sometime*. That is, although the likelihood of an accident occurring in Prince William Sound may be small, for instance, the likelihood of such an accident occurring sometime over the life of the tanker terminal is higher, and although the likelihood of anything going wrong with any particular aspect of a landfill site is small, the likelihood of something going wrong during its lifetime is higher. These are likelihoods based in part at least on our judgment that complexity increases the likelihood of failure.

We have two ways of proceeding in making proper decisions about such matters as where to site a landfill and where to site a tanker terminal. We may sever the link between the magnitude of a harm and its probability and take the magnitude to weigh heavily when it is great, even if not known to be likely or known to be unlikely, or we can preserve the linkage found in the common mode of decision making and *presume* that the likelihood of an accident occurring in any particular context is not itself negligible, certainly never so negligible as to make planning for one unnecessary, especially when the possible harm is of great magnitude.

We find ourselves calculating likelihoods on the basis of computer models of what-might-happen-when, and we have no way to resolve what computer model is itself most likely to make the most likely projections. The various suppositions about whether the earth is getting warmer and, if so, to what extent are classic examples of our lack of knowledge here.[64] But where knowledge is wanting, and we cannot fail to do something because even not acting is to continue current practices, and where our model for decision making requires an assessment of probabilities to act, we must act on some *presumption* or other about the likelihood of harm. The usual model effectively requires that, because of the harm that would result to current practices and interests were we to act without showing harm, we make a presumption of little likelihood of harm where the likelihood cannot be calculated: A projection of a high probability requires evidence. But, given our knowledge of how often human beings make mistakes, even in the most carefully controlled environments in which the aim is supposed to be to minimize mistakes—the *Challenger* disaster is a good example—we have, indeed, good reason to presume a relatively high likelihood of an accident's occurring, a high enough likelihood, certainly, to make planning for it a necessary part of any solution in which such an accident, of such a magnitude, *could* happen.

When we look back at the proposed hillside landfill site in Michigan, we have then two options. On the one hand, with the harm that such an occurrence *could* cause being quite significant, we need not assess the likelihood of such occurrences to raise the threshold. Everything else being equal, a different site is a distinct preference. On the other hand, given how many ways in which those siting and operating a landfill can fail to take due care and how many ways in which such a complex entity can itself fail, even with the most perspicuous foresight, it is not unreasonable to *presume* some degree of likelihood of an accident occurring over the lifetime of the landfill—sometime, that is.[65]

These two ways of proceeding thus have the same practical effect. They force us to treat worst-case scenarios as real possibilities. If we do, we would change the outcome of a decision process that treats the magnitude of possible harm as a negligible consideration even when great. That such decisions occur is unfortunately proven all too often by our discovery of their consequences, but we can also predict about some proposals, with a high degree of assurance given their details, that their consequences will be unfortunate.

For instance, the State of Michigan at one point agreed to provide for a low-level radiation dump site in which low-level radioactive waste, with a half-life of 5,000 years or so, would be put in containers or landfills with significantly shorter lives. We know that it is more than likely that someone will face difficult problems of not just disposing of the waste again but also transferring it in its leaky state. It is difficult to comprehend the reasoning behind such a proposal. It clearly does not treat a worst-case scenario as a real possibility to be avoided because it *invites* such a scenario.

One may only speculate here as to why a state would contemplate so acting as to condemn its future citizens (ignoring any harm that may occur to its present ones) to such radiation and its future leaders to such a problem. Clearly, the present political leaders will not be around to face the problem, and we may suspect the political incentives to dispose of the issue now, with the advantages up front and the disadvantages long down the road. In any event, although this case raises the issue of what our obligations are to future generations, if any, the possibility of harm to current residents from, for instance, too lax a standard for containing the radiation is enough to justify opting for an alternative solution, all other things being equal.

Again, to use a case I have remarked on before, it is unreasonable to dump steel barrels containing radioactive and toxic materials—plutonium (with a half-life of hundreds of thousands of years), cesium, and mercury—into the ocean, in an area known to be the most important bird-nesting site in North America outside Alaska, and known to be one of the most important areas for commercial fishing on the West Coast. Here the likelihood of the steel barrels remaining intact for the lifetime toxicity of its contents is nil: Steel rusts in water, and salt water corrodes. In fact, we know, without experimentation here, that it is likely that such barrels will rust and corrode within a relatively short time, twenty to fifty years. The potential harm to wildlife, the snappers, soles, herring, the seals and sea lions, the 300,000 sea birds, the humans who eat the fish, and thus the harm to those who depend on fishing and fish products is enormous.[66] That the area in which these barrels were dumped is within 50 miles of the San Andreas fault adds to the concerns.

We may again speculate on what those who approved the dumping were thinking, but it is hard to suppose that they seriously considered *either* the magnitude of the possible harm *or* that and the probability of the harm occurring. Had they considered either, they could not have reasonably done what they did. It takes no special evidence of how likely it is that such barrels will leak. Background information about putting steel in salt water for any extended period suffices to support a projection that the barrels will leak. Such a decision to dump steel barrels in salt water could only be reached if the likelihood of harm were severed from a consideration of the magnitude of harm and if *neither* weighed heavily in the decision making, an example of the poorest possible decision making.

It is, sadly, too easy to find similar examples of such decision making. Consider the nuclear waste at Hanford Reservation in Washington. The issues there are complex, and there are at least two separate problems.

First, there is the possibilily of a catastrophic explosion. "Although the risk analyses are crude, each sucessive review of the Hanford tanks indicates that the situation is a little worse."[67] The fuel from the nuclear reactors there was dissolved in acid so that the plutonium could be separated, to be used eventually in bombs. The acid-chemical mix that was left was put in waste tanks. It is radioactive because of the uranium and "combustible because of the chemicals added during the uranium separation process."[68] The mix was also acidic, and so it "had to be neutralized to avoid corrosion of the steel containers." The addition created a sludge in the bottom of the storage tanks, and then "organic chemicals were added to bind with the most radioactive material and make it sink to the bottom, creating a concentrated sludge." But all this has further concentrated the radiation into potentially explosive "hot spots," and the added chemicals themselves "break down into explosive or combustible compounds." So a mix that was already combustible and radioactive has been made apparently more so through efforts to control it.

The primary concern is of an explosion like the one that devastated and contaminated "thousands of square miles" in the southern Ural moutains in the Soviet Union in the late 1950s.[69] If we are not to assume conspiracies afoot or madness gone wild, the creation of waste tanks with such potential may best be described as an unintended effect of what must have seemed reasonable acts to control the mix.

Explaining the second problem without finding fault would require considerably more charity. It is a wonderful, though sad, illustration of the worst-possible decision making in which neither the magnitude of the potential harm nor its likelihood was taken into account. For the tanks that hold the explosive mix are of two types, both steel. The one sort is single-shelled, the other double-shelled. "In 1948, the Atomic Energy Commission said the single-shelled tanks are going to corrode in 25 years, and they were right, they did," said Leo P. Duffy, the offical in charge of waste management in the Bush administration. The consequence is that over one million gallons of the mix "have leaked from 66 of the tanks," and more is on the way.[70]

It is difficult to understand how anyone could have made such a decision to store such a mixture in such tanks unless compelled by something thought at the time to be considerably more pressing than concerns about the magnitude *and* likelihood of harm to the residents of that part of the State of Washington and to the workers at the Hanford Reservation.[71]

Radiation was leaked into the air from 1944 until 1971, and, in addition, because of Hanford, the Columbia River is "the most radioactive river in the world."[72] In the first place, "millions of gallons of water pumped directly through the reactor cores picked up enormous quantities of nuclear material," and in the second place, leakage from the tanks has presumptively entered underground water and made its way to the river.[73] The result is "the equivalent of 20 pounds of radium in the river every day." One result is that "in the early 1960's, a Hanford worker set off alarms when he walked through a radiation monitor at the plant" because "the day before he had eaten a can of oyster stew contaminated with radioactive zinc. The oysters had been harvested in Willapa Bay, along the Pacific Coast in Washington State, 25 miles north of Astoria."[74] The only known possible source of the contamination was Hanford.

The decision-procedure I recommend not only has the advantage of allowing us to avoid such disasters, but when the magnitude of possible harm is so great, it also gives us that advantage without requiring us to go into a black hole of calculating probabilities with too little evidence, as the standard mode of procedure presumes, and without requiring that we make predictions about probable harm, predictions that project themselves to be contested.[75] I see it as an added strength of the recommended decision-procedure that we can often arrive at the same conclusions by presuming the likelihood of an accident as we do by taking into account the magnitude of possible harm, even if unlikely. But it is more likely to affect practice in regard to at least some environmental issues, I suspect, to take the magnitude of harm as a weighty consideration without a consideration of the likelihood of the harm occurring.[76] Such a decision-procedure would make such decisions as those that led to siting an oil terminal in Prince William Sound or those that might lead to siting a landfill on a hill above a river much more difficult to sustain. We cannot play down the magnitude of harm because the likelihood of its occurring is small.

Yet, we have no formula, I want to emphasize again, into which we can feed problems to derive a solution. The nature of the subject matter does not admit of that sort of certainty. We do have clear cases where action is reasonable and clear cases where it is not, and we can compare new cases with our clear cases to come to some understanding of what is reasonable and what is not. With these in mind, we can examine our natural social artifact regarding waste.

The proposed change in our decision-procedure can be applied to particular issues within this artifact as well as the natural social artifact itself.

Either way, it ought to change the structure of the discussion about what we ought to do, and though we need to examine individual problems and assess the weights of the various variables to determine what we ought to do, there is a general cast of mind being suggested that gives a quite specific direction to our actions.

A Careful Cast of Mind: Easy Cases

An obvious easy case is lead. When ingested it causes lead poisoning. If it gets into our environment in a way that it can be ingested, then it will harm those who ingest it. It does get into our environment through, for example, old lead-based paints and batteries that are dumped or put in landfills so that the lead can leach into ground water. So our waste practices increase the likelihood of harm to our health.

This is not just an abstract possibility. A recent survey of a number of school districts shows that even after a strict law limiting the amount of lead in drinking water, findings "confirm that harmful amounts of lead exist in the drinking water provided by schools."[77] Lead *can* do damage to human health, and children and pregnant women are most likely to be damaged. The finding shows only that lead is in the drinking water of many school districts and thus that the likelihood of harm being done is higher than it would be if lead were not in the drinking water, not that any actual harm is being done.

How much harm is actually being done by there being lead in the drinking water in schools? It is unclear, but even if we do not know exactly what harm is being done to our health by all the various elements and compounds that are getting into our food and drink, for instance, it is reasonable to act in the face of the risk we face in not acting. For there are some things we can do, as with the CFC case, that will cause *comparatively* little harm to our interests. We can reduce the total amount of lead we release into our environment, for instance, and we can recover and recycle some at least of what we must produce.

In the face of doubt about what really are the results for our health of our present practices, but knowing that those practices increase the likelihood of great harm and may be causing harm, other variables in the situation become more important in directing our actions than our knowledge or our ignorance. The usual discussions of public policy issues tend to be driven by emphasizing our ignorance of possible harm and our knowledge of real costs. They ought to be driven more by emphasizing our ignorance of real costs and what knowledge we do have of the magnitude of possible harms. To be fully reasonable, our solution should be driven by all the relevant variables, properly arrayed for each particular case.

We have to go through each situation individually. It seems unlikely that a general rule exists by which we can weigh the various variables. But that does not mean that we are without any general guidance. In regard to waste, what is important is that we need not produce so much waste and need not throw away what we could use again. A careful cast of mind requires that we cut the sources of potential harm when we can, especially when such pruning causes comparatively little harm to our current practices. We can do much even without a showing of actual harm.

Consider the problems presented by heavy metals in our environment. The disadvantage of heavy metals from the point of view of disposal is that they cannot be incinerated. They remain in their elemental state and are either dispersed more widely if they are not captured in ash or collected in more toxic concentrations if they are—and they are—toxic. Mercury, for instance, causes severe damage to the neurological system. It has been added to latex paint to impede the growth of mildew and bacteria but can cause acrodynia, a mercury poisoning in children, marked by nerve disorders, leg cramps, loss of skin, and so on. That reasonable substitutes exist for mercury in paint is shown by the use of alternatives by major national paint producers. Mary Agoes, a medical epidemiologist for the Center for Disease Control's Center for Environmental Health and Injury Control, said, "There are other preservatives that can be added to paint that don't pose a health problem."[78] So we have, with mercury in paint, an element introduced into the environment that need not be used but that can cause severe harm. The matter became a matter of real concern when real harm was caused to a 4-year-old Michigan boy, hospitalized for four months after the interior of his home was painted "with the windows closed and the air-conditioner running," with paint that "contained mercury at a level above 900 parts per million, or about three times the legal amount [permitted to be] used by producers."[79] But we need not show actual harm to any person to be concerned. Mercury can cause harm. So having it in paint that is used extensively in painting interiors of houses where people will be exposed to it poses a high probability of harm to people. And alternatives readily exist. We ought thus to act to eliminate mercury from latex paint.

The recent reduction of mercury in batteries, remarked on earlier, is a good example of acting to reduce the total amount of potentially harmful waste we produce, with comparatively little harm to our interests. The new batteries being produced are as powerful as the old, and they contain significantly less mercury than the old ones, less than one-tenth of 1 percent of the total versus 7 percent of the total previously. Again, we need not show actual harm to act, given that change can occur with comparatively little harm to our actual practices, and there is the savings to these companies that comes from not having to purchase mercury and spend what is necessary to store it and dispose of whatever of its residue is left from the manufacturing process.

It is the cast of mind that matters here, the sense that we can do with something less likely to cause harm if we put our minds to it, at least in those cases in which no serious harm is done to our real interests. Ridding latex paint of mercury and reducing the mercury content in batteries can be done virtually without loss, and no one doubts the toxicity of mercury even without a finding of real harm to anyone. Other reductions in mercury may produce losses, some minor, some major, and the need for reduction may be disputed. We will have to assess each individually, but potential difficulties elsewhere do not change the appropriate cast of mind.

Consider another case where toxicity is high, but where there is some loss—the use of color in printing packages. Heavy metals are used in pigments to create bright colors: chromium and lead for yellow through orange, cadmium and mercury in deep bright red, and so on. These metals can have clear effects on our health: "Arsenic, cadmium, beryllium and lead are carcinogenic metals; arsenic, lead, vanadium, cadmium and mercury are neurotoxic; zinc, copper and mercury are acutely toxic to aquatic life."[80] If the colors in packaging are produced with such metals, they enter the environment in two ways. Incineration of the packaging releases them, concentrating them in the ash residue if they are captured, spreading them far and wide if they are not, and recycling the packaging releases them in the first stage of recycling when the material is pulped in water, and we end up with toxic waste water. Again, we need not show actual harm from the use of such pigments. The harm such metals *can* cause is well-known, and although there is some loss in their removal from pigments, the loss is small. "It may not be possible to have a General Motors blue or a Coca-Cola red."[81] In addition, except for lead chromate yellow, which has proved hard to replace, the difficult elements can be replaced by carbon-based colors, and "carbon-based inks will break down in incinerators."[82] So the most we will lose is some brightness in packaging. If General Motors does not survive, it will not be because it lost its blue.

We can see the same careful cast of mind at work in downsize packaging, a nice illustration of how a convergence of interests can occur to the benefit of the environment and of how the proper cast of mind can be used regarding cases in which the magnitude of potential harm is not so great. Detergents, to take one example, have for years contained fillers such as sodium carbonate. These are not essential to cleaning, but, although they somewhat ease production, they are primarily used to make consumers feel they are getting more for their money. New superconcentrated detergents, without so much of the fillers, can be packaged in much smaller containers than the old jumbo boxes. The amount of packaging that may be covered with advertising and thus with pigments made with heavy metals is that much less, but the real gain is that less packaging material is needed and less waste is left that must be recycled or disposed of in landfills or by incineration. Because the new detergents are less expensive to produce, and because the packaging takes up less shelf space that supermarkets want so they can sell other items, the switch from jumbo to superconcentrated boxes serves well a variety of interests, including an interest of manufacturers to save money in packaging materials. Of course, consumers need to be convinced that less is more,[83] and detergent manufacturers are still responsive to the presumption that more is more. "Cheer is concentrated so that half a cup—not a quarter or a third—equals a full cup of traditional powder."[84]

It may be objected that the cases considered so far are relatively easy: The potential harms are relatively clear, the solutions readily available, and the harm to existing interests caused by change relatively small or nonexistent. But part of the point of providing such examples is to show that much reduction of waste *is* relatively easy. We need only put our minds to it, and, with careful consideration, reduce where we can with a minimum of losses. Yet, the careful cast of mind I am recommending and the decision-procedure I propose does have a cutting edge—as my remarks on CFCs and on siting the oil tanker terminal in Prince William Sound illustrate and as can also be readily illustrated by the difficulties of dealing with the predicted greenhouse effect.

The Bush administration was unwilling to consider any significant action to diminish the incidence of those gases thought crucial to the greenhouse effect. The argument turns on the claim that no one "should be asked to spend money to solve an environmental problem that may never materialize. How much," after all, "should be spent on a greenhouse solution before it is certain there is a greenhouse problem?"[85] Passing regulations that would force companies to reduce their relevant emissions would cost them money, and so harm them, and that is unreasonable, so the argument continues, without a showing of actual harm.

Yet, because the magnitude of harm that may occur is so great, it is not unreasonable, I am arguing, to act in ways that would diminish the harm *if* we can do so with comparatively little harm to our ongoing interests. A properly cautious cast of mind would take seriously the *possibility* of harm of such great magnitude as would occur if the earth's atmosphere does warm as much as it has been predicted it may, just as a properly cautious cast of mind takes seriously the possibility of the loss of the ozone layer. The appropriate action then, it has been claimed, is to "buy some insurance against catastrophe, and buy it as cheaply as possible."[86] The insurance we buy is not to allow us to continue what we have been doing, risking a catastrophe, but to diminish the likelihood of such a catastrophe by acting to minimize it at minimal expense.[87] This cast of mind, which acts to reduce the harm of potential risks when the magnitude of that potential harm is so great, is strikingly at odds with a cast of mind that rejects all attempts at any reduction because there is no clear evidence that the potential disaster may occur.

Even if our epistemic position is shaky, even if we really do not know what the likelihood of harm is or, indeed, whether there is any likely harm at all, then the magnitude of potential harm still is enough to justify action, but action properly tempered by our epistemic shortfall and by the harm that will occur to existing interests. So it is a reasonable response to encourage a decrease in the emission of greenhouse gases, and that is a response made more reasonable by its serving such other interests as the conservation of energy. We should thus act to do what we can to cut emissions worldwide,[88] and we should act to do what we can to cut emissions in the United States. And those acts that have priority would be those that are least expensive or most helpful to other interests we have.[89]

"So," one hears someone asking, "how much insurance do you want to purchase? And at whose expense?" In short, the objection to acting now to reduce likely causes of the greenhouse effect is not just that there is no showing of harm yet but also that, when one must act, one must *do* something and thus harm actual interests and must harm actual interests without showing that those are the ones at fault or most at fault. We must commit ourselves to reduce emissions, for instance, and at a cost.

Yet, it seems perfectly reasonable that we ought to act to forestall a potentially catastrophic harm even without showing actual harm, or even much likelihood of harm, when we can act in ways to reduce the risk and yet cause *comparatively* little harm to our interests, if any at all. I have concentrated on heavy metals, for instance, because no one doubts they cause harm and because, as it turns out, their reduction in the environment in some clear and significant cases can occur in a way that benefits those corporations that use them. It is not insignificant, for instance, to reduce even by a small fraction the amount of mercury found in batteries. "Americans now consume 2.7 billion batteries . . . a year," and they "account for 52 percent of the cadmium and 88 percent of the mercury found in trash."[90] But once we shift our cast of mind and begin to search for ways to reduce the production and harms of waste, we find ingenuity at work to use enzymes that readily break down once they are used to break down stains in clothes,[91] soybean oil to produce brighter colors and blacker blacks in newspaper ink,[92] cloud gel between window panes that lets in sun when the gel is cold and keeps out the sun when the gel is warm so that heating and cooling costs are lowered,[93] an inexpensive method of converting "used motor oil and other lubricants into gasoline and heating oil,"[94] an enzyme to "rid underground drinking water of nitrates, a dangerous and increasingly widespread pollutant,"[95] a method for removing "virtually all the dangerous compounds in fumes produced by chemical processes,"[96] the creation of a new fuel based on canola oil for diesels that "does not release sulfur dioxide" and "reduces emissions of other pollutants by 50 percent,"[97] and so on.[98] In short, once we change our cast of mind and strive to find ways to reduce the risk, we shall find that many things can be done relatively inexpensively.

Indeed, I would argue that our current natural social artifact of waste management is essentially incoherent. On the one hand, it reflects a way of looking at the earth and our place in it that encourages our producing what we need without much concern about environmental costs or waste. It certainly does not encourage the reduction of waste. It is also, on the other hand, driven by a decision-procedure that encourages leaving things as they are until it is shown that real harm occurs. As I have argued, incineration thus emerges as the preferred option. We are least likely to cause harm ourselves by leaving as much alone as we can, and incineration, like landfilling, has the virtue of allowing us to leave things as they are. We continue to produce what we need without much concern about waste, and we just burn instead of burying.

Yet, incineration itself causes harm. First, it does nothing to encourage the reduction of harmful elements, such as mercury, from the waste stream by discouraging their use in such products as batteries. To the extent that we have in place a way of handling waste that does nothing to discourage the introduction of harmful products in the environment, we have in a place a way of handling waste that contributes to our being harmed.

Second, it encourages the production of waste rather than discouraging it because once an incinerator is built, we need waste to keep it going at a profitable level.[99] It thus not only does nothing to discourage the production of products with harmful elements in them such as mercury but also encourages at least enough continued production of waste to ensure the incinerators a sufficient supply of trash to burn.

So we have a natural social artifact that produces incineration as the preferred solution to the waste it produces and so further entrenches the harm created by the artifact. But it is essentially incoherent to refuse to act to prevent possible catastrophic harm from occurring and then act in such a way that will further entrench existing and known harms and keep them from decreasing.

This is especially so when the change in one's cast of mind can so readily produce such quick savings as are illustrated by what has happened to mercury in batteries or by what we have discovered about how we can clean circuit boards without CFCs.[100] The conclusion to be drawn is thus that we have our priorities backward. We ought to put as our first priority the reduction of waste and, in particular, the reduction of what is harmful in the waste that we produce. That reduction has two components. We need to reduce what enters the waste stream, and we need to reduce what we need to get rid of as waste. The former encourages the sorts of profitable reconfigurations of products and processes I have illustrated with mercury and circuit boards. The latter encourages recycling. Both forms of reduction can be achieved in many cases with comparatively little harm to our interests.

The Relevance of Rights

Reducing the amount of waste we produce and recycling what we can will not rid us of problems. We will still have to deal with what we already have and what we will produce that we are unable to recycle. But the argument does not propose that we do nothing but reduce waste and recycle. It is meant to show that waste reduction and recycling ought to be the dominant components of any comprehensive long-term solution to the problem. In particular, we ought not to wait for a showing of actual harm to act. The magnitude of potential harm is sufficient to justify action, provided that other relevant variables have enough weight.

Of course, again, one might object that they do not, that, in particular, the interests we have in the way we now produce and manufacture goods and in our particular habits of use would be harmed if we reduced our waste production and recycled what we can. If so, then, it would be argued, we would not have shown that waste reduction and recycling are *better enough* to justify the disruption to our current practices they would entail.

It looks, in other words, as though we are only weighing interests against one another just as the political process tends to force us to do. If we are just weighing interests, then it is not *obvious*, it may be claimed, that waste reduction and recycling meet our interests better, especially in the harder cases. The balancing of interests one against another is a difficult business in any case, and where we are unsure what interests are being protected but are sure what interests in our current practices are being harmed, then, it can be claimed, it is certainly not obvious that we ought to act when we know that by acting we will cause harm.

In fact, however, we are really not weighing just interests against interests, because on neither side is it true that just interests are involved. On the one hand, we have an interest that our health not be harmed by the introduction into our environment of noxious material. This interest is of such importance that it is arguably a *moral right* that needs to be weighed in. If this right is denied, then we have been wronged,[101] and we thus have two weighty reasons for acting. When a right is denied and our interests set back, we have been both wronged and harmed.

On the other side there is a lack of a moral right of manufacturers, among others, to use common resources without paying for the cost of their disposal. So no moral rights of theirs are being harmed by preventing them from producing waste they do not dispose of. If we now weigh in a moral right on one side of the scale and the lack of a moral right on the other, the balance is tipped in favor of waste reduction and recycling, even if we add in the disruption to existing interests that would result, including unemployment.

Not everything introduced into the environment is harmful, but when something is introduced that has been shown to be harmful or shown to increase the risk of harm, then it becomes a *presumptive* target. That is, it ought to be removed or mitigated unless there are good reasons for not removing or mitigating it, and the good reasons have to be good *moral* reasons, based on competing moral principles that are weightier in the context than our right not to have our health harmed, if we have such a right. One must argue case by case.

The form of any argument is like the form of arguments weighing interests. Just as slight damage to some of our interests is outweighed by heavy gains to others, so some wrongs can be outweighed by heavy gains regarding other rights. We may commit ourselves to a public transportation system that damages existing rights, to property, for instance, because the gain is an increase in real equality of opportunity for those who need public transportation to obtain and travel to jobs. But although the form of the argument is the same, its elevation into the realm of rights means that we must find competing rights to weigh, for instance, against the right not to have our health harmed by the introduction of noxious elements into our environment. The ante has been raised by requiring that it be shown that not just interests have been harmed but that rights have been damaged.

But do we have a moral right not to have our health harmed by a noxious environment? What justifies treating as a right our interest in that? A thorough answer to this question would take us far afield into questions about the status of rights, how interests become rights, and so on.[102] But we can get a sense of how any such extended argument must go by considering what kind of interest it would be if it were just an interest. For clearly it is not just one among the many that each of us has, for so many of our other interests depend on the satisfaction of our interest in not having our health harmed. It is hardly worth my while, for instance, to take an interest in taking photographs to show my friends and relations, in ordering books to read some time in the future, or in committing myself to writing papers for conferences two or three years from now—all interests I have and activities I regularly engage in—if my health deteriorates significantly enough to make these activities difficult or impossible. Many of our interests look to the future in a way that make them depend on our having our reasonably good health in the future, and so, in that way, the interest we have in not having our health harmed in any serious way is a condition for many of our other interests. In that way it would hold a special place among interests. It is a welfare interest, and if it is set back, a person is seriously harmed, for setting it back sets back as well all the interests that depend upon its realization.

The same sort of argument can be made about its status vis-à-vis what we clearly recognize as rights. We each have a political and moral right to equality of opportunity, for instance, a right that looks to the future in presupposing that those who have it are seeking opportunities for advancement—in education, in housing, in employment, among other goods. With one's health harmed, one's future prospects are diminished at least to the extent of the harm, and, thus, such a right as equality of opportunity has that much less value to someone whose future is diminished either because the person will die or will live substantially impaired.

Such considerations suggest that it is a moral right we have, not just an interest, not to have our health harmed by noxious materials in our environment, and we can reach that conclusion in another way as well. If we consider the ends of political society—why we organize society one way rather than another, what we can possibly intend in preferring a democracy, for example, over a tyranny—then we are led to the claim that only in a democracy can citizens have a chance to develop themselves fully. Only in a democracy can they be free to explore the intellectual world, to see what talents they possess and press them to their fullest, to discuss and argue and engage each other. Underlying our preference for democracy, in short, making it a principled choice and not just a preference, is a commitment to the individual and to the value of each person. That commitment ultimately underlies the claim that not having our health harmed is a right. For only if the health of individuals is not harmed can we hope, in a democracy, that each individual can develop himself or herself fully.

To develop these arguments fully would take us too far astray, but their success for our purposes here depends on another claim's being true as well, that no one has a right that outweighs generally our prima facie right not to have our health harmed. For the right that our health not be harmed by a noxious environment is a prima facie right, mediated by a consideration of other rights that may conflict. It is not absolute, that is, but a right that can be outweighed by other considerations, though (presumptively) only if those considerations are moral.

If manufacturers of toxic substances had a right to manufacture their goods and to pollute the air and water in the process, then if that right were more weighty than our right not to have our health harmed, the right would not be of use in the present situation. But they have no such rights. Again, the argument here could easily take us far afield into questions about how, for instance, we weigh rights one against another,[103] but we need not settle that issue, and the two parts of the argument we need can be quickly sketched.

First, each manufacturer, mining company, oil producer, or whatever is a corporation licensed by the state or states in which it operates, and nothing about such licensing prevents a state from restricting the corporations in various ways. In particular, nothing prevents any state from prohibiting companies from producing waste they do not dispose of properly. The proper comparison here is with licenses to professionals to practice. These are privileges awarded by the state to those who have been judged competent in particular areas of expertise,[104] and nothing prohibits states from conditioning the awarding of such privileges. A state might require a lawyer to do so much pro bono work, for instance, in order to be licensed to practice within its borders. Just so, a state may condition a corporation's manufacturing within its borders on the corporation's properly disposing of its waste, or, more reasonably, to ensure that no company in any state is at a competitive disadvantage, a national law may require that.[105]

If corporations are not required to dispose of their own wastes, the wastes become a public problem, a disposal issue that is paid for by the taxpayer either directly, in the form of taxes to pay what the corporations would otherwise pay, or indirectly, in the form of costs to health and the environment that degrade the taxpayer's quality of life. It is a political question, to be decided by a state, whether a corporation pays for the disposal and passes on the charges to consumers of its products, or whether it does not pay for the disposal and the consumers thus pay for it directly or indirectly. Nothing about the nature of a corporation, of a state, or of waste prevents a state from requiring those who produce waste to dispose of it properly.

Second, much that is used in the manufacturing process is common property—air used in foundries, water used for cooling purposes, and so on. In addition, it is dumped either immediately or circuitously into common property—into the air or the water. Some of what is used is by its very nature reused: Water and air are recycled. But because they are common property and do not belong to those who use them in their manufacturing processes, manufacturers have no right to degrade them, and because air and water are of such vital concern to our health, we have a right that they be returned to us relatively unpolluted by their use, and at least fit for reuse.

We have had a long history in this country of treating air and water differently. Both have been treated as common property insofar as dumping is concerned, but air is of such abundance and the needs for its use so independent of its quality that no manufacturer needed to claim ownership to guarantee an adequate supply for smelting, for instance. Even if the air is polluted by fumes, it is there to be used, and no manufacturer generally needs to act to limit its use by others to have a sufficient supply of the quality of air needed for manufacture.[106] But manufacturers have wanted to treat water as private property to guarantee a supply. They have wanted to treat a more scarce common resource as their property, excluding others from using it to guarantee that they have full access to it. We need only imagine old mills dependent on a constant stream of water to supply the power necessary for their work to get a sense of how important it is for some manufacturers to make sure that others, upstream, not diminish the supply of water available.

The argument for treating water and air alike as common property depends on our recognition that our water and air supplies form a constant recycling system and that the degradation of it at any point may affect its overall quality and so impinge on our right not to have our health harmed. One objection to incineration is that "it turns the atmosphere into an uncontrolled waste dump for toxic pollutants like lead, mercury and dioxin."[107] We have no trouble thinking of a watershed. The concept allows us to concern ourselves with how what degrades one part of a watershed has effects on other parts. Similarly, we should think of our air as an airshed, a concept that opens space for us to consider how air pollution in one place affects us elsewhere.

If the claim of our right that air and water not be degraded by their use is not obvious, that is only because we have for so long allowed free use and could do so because natural processes cleansed the air and water for us and we have had an abundance of both.[108] If we did not have such a history of inattention to our common resources, or if they were not so common as to be almost invisible as a resource, it would never occur to us to suppose that, without paying for it, someone could use for profit something that belongs to all of us.

We have public parks, and we would be surprised at a sports team owner arguing that because a park is public property, the team has a right to use it without paying and could even charge citizens an entry fee to see the team play. We might allow for a fee's being charged, but generally only if the team paid for using the park. They have no right to use it, although we could, if we wished, give them the use free if the public benefitted in some sufficient way. It is because the public does benefit that an argument can be made that we allow city league teams free access to public parks.

If the analogy with what is clearly public property seems inappropriate, imagine a manufacturing concern in a city producing noise and vibrations by smashing cars with an hydraulic press or think of our regulating, as we do, the approach of airplanes over inhabited areas to prevent their bothersome noise during the night. We have no compunction in regulating noise to maintain relative quiet.

A state could decide that manufacturers are to have free use of such public and common resources as air and water, just as it could decide not to regulate the noise of manufacturing or of airplanes, but that is a political decision. Nothing essential requires that, and manufacturers might just as well be required to pay for what they use—in the present case, by returning to its original state the water and air they use and even, if they cannot do that, being prevented from using it or, if the product is too important, compensating somehow for the necessary degradation.[109]

The form of this argument can equally well be applied to the hazardous waste that is the bane of modern society. If a chemical company produces toxic by-products, nothing about the nature of corporations or states prohibits a state from requiring such companies to dispose properly of their own waste, including the funding of a state agency to oversee the disposal. Because such waste otherwise gets into our environment to put our health at risk, corporations not only have no moral right to produce toxic waste without disposing of it properly but also are morally obligated to dispose of it properly.

This moral obligation is, again, prima facie. That is, it is a *presumptive* obligation that can be outweighed, but only by competing and weightier moral obligations. The obligation is modified when proper competitors enter the field. So if a company produces harmful by-products while producing something essential to our health—radioactive material used for tracing in X rays, for example—and cannot dispose of those by-products properly because no safe disposal is yet available, it would clearly be morally wrong to penalize the company for failure. The best that we could do would be to demand that the company meet whatever conditions are deemed adequate to handle the waste until proper disposal is possible. Again, if the state has permitted a company to operate in the past without adequate safeguards for the environment, it is arguable that the state has thereby assumed some moral liability to clean up the environment, requiring public funding for what have become, through past state inaction, at least in part public wastes.[110] Sorting out the details of moral liability is a complicated business, and here, as elsewhere, details matter. But none of those complications change the prima facie obligation of companies properly to dispose of what harmful wastes they produce.

We may argue here on two different grounds to the same conclusion. First, it is both more efficient and more practical to require those who manufacture waste and so can identify it and control it to dispose of it rather than let it leak into the environment and out of any responsible control. Second, those who put the public's health at risk are morally obligated to mitigate the risk. Both practical prudence and morality argue for the same conclusion.

The same argument can also be used to justify control not just of the toxic wastes produced by manufacturers but also of the waste produced by their products. A manufacturer has no right to produce anything at all packaged any way at all. No manufacturer has a right to produce a dangerous substance that maims and to package it as a present to entice children. We would be appalled if anyone did that and more appalled if they argued that it was permissible to do that. Yet, if we can control what is produced and how it is packaged, we can prohibit bottlers from producing cans with tear-off tabs or producers of aerosol sprays from using CFCs. We could thus, if we had the political will, require bottlers to use bottles that can be recycled, as some states now do, or require that those who sell products in containers that are not biodegradable provide either biodegradable packaging or the means to recycle what would otherwise litter our environment.[111]

Such requirements would have vast effects on manufacturing. If we were to calculate into the cost of products the cost of the degradation of the environment caused by their production, just that portion of the costs, for instance, of building and maintaining water treatment plants needed to remove the impurities caused by production and use of products that require water or make their way into our water, we would fundamentally alter the way in which products are priced.[112] The styrofoam containers in which hamburgers are often packaged might well cost more than the hamburgers themselves. Clearly for products containing toxic material, mercury batteries, for instance, requiring manufacturers to bear the costs of handling the toxic material their products introduce into the environment would increase the costs of production.[113] Presumably, such costs would be passed on to the consumer, and the transfer of those costs would both lower the costs we now incur for the cleanup of toxic wastes and effectively change our habits of consumption. If the products became too expensive, we would buy others if we could or buy less, and if some products that were necessities became too expensive, such as needed medicines whose production causes harmful byproducts, we would have to consider subsidizing their costs. After all, recalculating the costs of products and pricing them accordingly would cause the products to be distributed among consumers differently, raising issues of justice regarding some products at least.[114] In any event, if we were to calculate the true costs of producing commodities in not further degrading the environment, that would affect our patterns of consumption.

It is unrealistically optimistic to suppose that we each will voluntarily change our habits of consumption for the sake of long-term goals that are not immediately foreseeable and whose realization depends, in any event, on others sacrificing as well. But if the political will could be exercised to require manufacturers to change their practices of production, our practices of consumption would be changed effectively just as well. It requires optimism, but not unrealistic optimism, to suppose that enough voters would act to tip the balance in interest calculations in favor of changing such manufacturing practices.

We can now return to the objection with which we began this section, namely, that if we are just balancing our interest in not having our health harmed against our interests in maintaining our present practices, it is not *obvious* that we ought to act to protect our interest in not having our health harmed. But, as we have seen, the argument is not so simple as that. When we identify the effects of particular practices, evaluate what they do to our interests, and so on, part of what we shall have to collate into the process are the various moral rights, lack of rights, and obligations. If our interest in not having our health harmed is a moral right, if manufacturers have no moral right to produce in any manner they see fit, and if they in fact have a moral obligation not to put our health at risk, then the balance surely tips in favor of waste reduction and recycling. It becomes obvious that we ought to do what we can to avoid the problems we have by reducing the extent of the problem.

Two different lines of thought take us to the same point. First, if we back off from our immediate problem of what we are to do with what we are now producing and look at our entire natural social artifact of producing waste and managing it, and if we note how much waste we produce and mark its ascending curve, it is obvious that we are overwhelmed, and will be even more overwhelmed, with waste we do not need to produce. Prudence dictates attempting to diminish the supply of what we have to manage just so we can begin to get a handle on it.[115]

Second, when we note the risk of harm to our health created by our current natural social artifact, the extent of that harm, ranging as it does from our water supplies to even our roadways and lawns and seashores, and the continuing and persistent nature of the harm, both practical prudence and morality dictate attempting to diminish the supply of waste we produce and recycling whatever we can. The situation we face is closer to that we face with CFCs than that we faced with PBB: Action is reasonable, and waste reduction and recycling are better alternatives. They ought to be the dominant parts of any long-range attempt to handle waste.

The most difficult aspect of making primary this prescription to reduce waste and recycle is that its implementation will certainly harm current interests. Focusing on our right not to have our health harmed and pointing out that a right ideally trumps an interest may look ideological and heavy-handed and deny the power of those current interests. That may seem to leave those with interests in our current practices with little to say in their defense, but, I am convinced, we need not worry that entrenched interests will not have enough to say. The system's penchant to transform all issues into competitions of interest, the apparently rational demand that we not harm current interests without all relevant evidence of their harmful effects, and the realization of how the economic consequences of harming those interests will severely harm the political constituencies of the political bodies that must make the political decisions, all these, I think, will conspire to give vested interests more than a fair share in any hearing on waste. Emphasizing that it is morally wrong to harm people's health is a proper inoculation against the systemic effects of the political and economic biases that a mistaken sense of rationality supports. The weight of that moral claim and a properly cautious cast of mind make reduction the dominant theme among those options we face.

Other Rights and Interests

We can reach this conclusion without any consideration of any other rights or interests we have, besides those regarding our health, that may be affected by our natural social artifact regarding waste. I do not intend to canvass them all or even examine the effects of any other. The structure of the argument is the same, although the weight of our rights and interests may vary from case to case. Yet, one other is of especial importance and needs to be noted explicitly, our complex interest in the environment—an interest in wildlife not being harmed, in the loss of possible benefits resulting from environmental damage, and an aesthetic interest in a loss of natural beauty.

Considering this complex set of interests is in one way redundant. Concern about our health is sufficient to justify action. Yet, these interests in the environment may justify particular actions and measures not justified by an appeal to our health, and, in any event, adding these interests to our list of concerns can only add weight on the side of waste reduction and recycling.

Each of these interests has weight, but complex questions that admit of much dispute and no great consensus infest them. Do we have merely an interest in the beauty of nature, or is there a right? If it is a right, is it like our right not to have our health harmed, or is it a different kind of right? Is it a right only we who are now alive have, or is it also a right of those who will follow us to inherit an earth whose natural beauty has not been maimed? Do animals have merely an interest in not dying unnecessarily or having their health harmed, or is it a right? Either way, is it an interest or right they have, or is it an interest or right we humans have for them or in them? Is it enough to argue for possible benefits that might accrue to us, such as benefits from various species of as-yet-unexamined plants, or must we show actual benefits that have accrued, and, in either case, what sorts of benefits, other than aesthetic or those relating to our health, are relevant? These questions do not admit of easy answers and may not even be the right questions to ask. We can, however, bypass the difficulties they raise.

Birds and mammals are undoubtedly being harmed by our capacity to pollute rivers, lakes, and seas to the point where aquatic life can no longer be sustained. Whether *they* have a right not to be harmed in that way, *we* certainly have some interest in not harming them. One may put this in purely economic terms. The pollution of the Chesapeake Bay has caused damage to the oyster beds and thus economic loss to those whose livelihood depends on oysters. In short, whatever the answers to the questions regarding whether animals have rights or whether aesthetic considerations justify rights or only interests, it is *sufficient* that *our* interests are being harmed and, in particular, that economic interests are being harmed. To give some industries free use of such public goods as air and water is to harm other industries—tourism that depends on fishing, clear air, and unpolluted trails, logging that depends on trees not dying prematurely from acid rain, oystering that depends on oysters themselves having a quality of life that guarantees survival.

No doubt vested interests would be harmed by cleaning up Chesapeake Bay, but it carries little weight in such a situation to argue, for instance, that some workers may lose their jobs. That is not obvious, for one thing, because the pollution may be prevented without closing any plants. We cannot tell a priori. For another, new jobs will be created by making a commitment to cleanse the Bay, and other new jobs will be created in manufacturing plants because stricter pollution controls will require personnel with new sorts of expertise. But the appeal that some workers may now lose their jobs rings hollow when one remembers how many oystermen used to work the Bay waters and how many have lost their jobs as the slow degradation of the Bay has forced more and more off its waters.[116] If we could factor in the economic harm already done to them,[117] the special pleading of those whose present interests may be affected becomes obvious. Indeed, even if some manufacturing jobs were lost, other jobs would be gained, as oystermen would be able to pursue their livelihoods once more.[118]

Putting it that way, as one set of economic interests against others, and noting our right not to have our health harmed, we have no compelling economic reason not to change our natural social artifact so as, at least, not to degrade our environment any farther. We could proceed to appeal to our aesthetic interest or right in having our environment unsullied, but I am convinced such arguments do not fare well among those we elect and charge to make decisions about what ought to be done. It is of little use to give arguments regarding public policy matters that do not persuade. Just as aesthetic considerations do not fare well against entrenched economic and political interests in arguing for the preservation of historic buildings, so that we best argue that it pays, economically, to preserve them, so I think it is best, because more persuasive, to argue on economic grounds for the preservation of our environment.

I want to bypass complex and contentious moral questions about animal rights, for instance, by appealing to an interest dear to those whose economic interests will be harmed by any attempt to reduce waste and recycle what we do produce. If the degradation of our environment is a trade-off of economic interests and is accompanied by harm to our right not to have our health harmed, then those who argue that there will be an economic loss are revealed as arguing that *they* will lose economically. The economic argument, which has the appearance of objectivity, becomes self-serving.[119] Although they may lose economically, others will gain.

How to Proceed

In turning from interests to rights, we find ourselves back again comparing interests. It may be that the mode of thought that so dominates political disputes has infected us so deeply that we are unable to get out from under it, but, whatever the cause, as that mode of thought so dominates political disputes, we would best operate by making use of it. If we can argue for a conclusion using that dominant mode of thought, that is one less variable we need worry about having to counter. Indeed, appealing to interests rather than rights has the virtue of appealing to weaker premises. If there are relevant rights, so much the better for a conclusion we can derive from an appeal to interests.

An appeal to either will change the way decisions are made about a variety of environmental issues and about waste in particular, given a change in the way in which we think about such issues. The details will vary from case to case, and because I think they can matter enormously, I do not want to lay down some detailed framework within which everything must fit, come what may. It will be much easier to do something to ward off a potentially disastrous harm if there are things to do that are effective and cost little than if there are not. The calculation about what we ought to do may well turn on such details.

But the general framework is in place. The vision that lies behind many of the debates about how to respond to environmental issues, and to waste in particular, is that full information is required for a fully rational decision. This is a vision for what it is to know that one is doing what is rational. But when that vision is combined with a concern not to cause unnecessary harm to existing interests, it produces inaction and privileges the status quo, even if continuing the status quo may produce a major disaster. But, I have argued, it is perfectly reasonable to make decisions without full information. We do it all the time and are reasonable in doing so. And it is perfectly reasonable to respond to potential harm when the harm is of great magnitude even if it is unlikely or not known to be likely. Again, we do that all the time and do it appropriately.

It is thus perfectly reasonable to respond to the threat posed by CFCs by ceasing the manufacture of CFCs. It is perfectly reasonable to respond to the threat posed by greenhouse gases by working to diminish the production of the gases that create it. It is perfectly reasonable to attempt to ensure that our water supplies do not deliver lead to our homes.

If we concentrate on our waste, we can see that it is perfectly reasonable to work to diminish heavy metals from the waste stream because of the harm they may cause in the environment or to provide the sorts of incentives that were in vogue for alternative forms of energy reduction by providing economic incentives for reducing and recycling waste. These can take a variety of forms. They can range from providing such seemingly minor, but effective, incentives as a small return on bottles for recycling to heavy taxation on those industries that fail to be responsible for their own waste. The former provides incentives for consumers, the latter for producers. We can alter the incentives for utilities so that they are encouraged to conserve energy rather than sell more,[120] and, in concert with that, we can do much more than we are presently doing to encourage efficient use of resources.[121] We can provide minor incentives for a wide variety of products that we do not want dumped into the environment and that can be recycled—from lead batteries to tires. Municipalities can charge households for the amount of trash they actually discard, charging by the bag, for instance, and thus encouraging citizens to recycle and compost.[122]

We may wish to go so far as to criminalize the conduct of those companies which put the health of their employees, their customers, and the general public at risk or even to prohibit the operation of such companies that cannot, or will not, provide adequate measures for the disposal of the waste they produce.[123] We certainly would want to go so far as to provide economic incentives for them to reduce the amount of waste created by their products and to make even that easy to recycle.

Each of these responses will cause harm to existing interests. So it is more reasonable to opt for solutions that least harm those interests, but, again, that is something we do all the time. When the barbeque flares, we do not dump buckets of water on it, but spray it with water to dampen the flames while keeping the charcoal hot. The cast of mind I am arguing we should adopt for environmental issues and for waste management is one we are all familiar with, find perfectly appropriate in other contexts, and is perfectly appropriate for environmental matters. What we ought to do is to take that cast of mind and embed it in our way of thinking about environmental issues, and of waste in particular.

CHAPTER SEVEN

A New Cast of Mind

On a local call-in talk show about the greenhouse effect, a caller asked one of the guests whether, given the various uncertainties about the matter, it would not be wisest to act now to minimize the possible harm, at least spending the minimum to discourage emissions of the relevant gases. The longer we put off any action, the worse any problem might become. More harm would be caused, and it would be that much harder to reverse the process. The guest responded with the cast of mind I have been castigating here. He said that such a suggestion was about the most unscientific thing he could imagine because it was based not on any evidence of harm but only on the suspicion of possible harm. The word "unscientific" was meant, by the tone of voice, to carry just the pejorative ring we might suppose it has. For "unscientific," read "unreasonable."

It is against the background of that cast of mind that what I have to say must be read. Nothing I have said will, I hope, be read as implying or suggesting that that cast of mind is somehow just wrong or inappropriate. Ensuring that decisions are made only with full information is an heuristic ideal that ought to inform scientific enquiry, and we have all found ourselves in situations where we have castigated ourselves for acting, when we need not have acted, without sufficient information. Failing to read the directions thoroughly on a new alarm clock and so setting it for P.M. instead of A.M. is, as I have suggested, a simple example of an all too common practice.

Yet, when that cast of mind is transported to public policy matters, even those regarding environmental issues that require an enormous amount of scientific expertise to understand, it can become unreasonable, despite what the talk show guest suggested. This is especially so when a failure to act may well itself cause harm. For if we approach issues of public policy with that cast of mind, issues such as those regarding waste management or the greenhouse effect, it will privilege the pre-existing natural social artifact because it will be unreasonable to do anything without full evidence. If, however, in response to the charge that we are unreasonable not to get all the facts, we get caught up in the discussion about what is known and what is not, we shall find ourselves inevitably with an epistemic shortfall—sucked into a black hole of claim and counterclaim, of calculating probabilities and improbabilities with too little evidence, and unable to make a decision because we do not have all the evidence. If we adopt the usual mode of decision making, someone can always and rightly play the epistemological card, for we can never know all that we ought to know to be sure that we are making what is, by that model of decision making, a fully rational decision.

You will find that even if you succeed in preparing a good case that great harm may occur from some proposed course of action, such as siting an oil tanker terminal on a pristine and protected sound, you will be met with the response, "But it is so unlikely!" If you dispute that claim, you will enter another black hole of claim and counterclaim, with similar results. If you do not dispute it and allow it to stand, that will be the end of the matter, thus privileging the existing natural social artifact. That you must either dispute the claim or not is, however, a false dilemma, a dilemma you face if you accept the usual model of decision making. For by that model you must *know* what the real likelihood of the projected harm is. The exclamation "But it is so unlikely!" is backed by the apparently quite reasonable concern that we not protect ourselves against unlikely prospects when doing so involves known harm to existing interests.

It is a difficult political or economic argument to make that we should cause immediate harm because of possible harm, and it is much more difficult to make it when the model of decision making adopted makes it unreasonable to act without full information and makes it especially unreasonable to act, without full information, in response to an unlikely eventuality. Anyone suggesting protective moves, such as the caller in that talk show, will be seen by anyone holding that natural and compelling model of decision making as perhaps slightly out of it and certainly unrealistic—suggesting something unreasonable, impolitic, and uneconomical.

It is no accident, in response to this model of decision making, that I have concentrated on examples such as heavy metals, lead in our water supply, chlorofluorocarbons (CFCs), or gases that may produce a greenhouse effect. Each of these cases is different, but in each case we can take action now that is reasonable: We can mitigate *present* harm as well as potential future harm. These are the easy cases, and the variations among them primarily turn on how much current harm is being mitigated, on how expensive it is to act now, and on how much current interests need to be impacted by what we do to mitigate both present harm and potential future harm. Again, the clearest examples for my purpose are those in which acting now produces clear benefits, as it does with getting rid of new sources of lead in the water supply, without much harm to our current interests. It is not necessary that lead be a component of any new water faucets, for instance. We can readily provide examples where the cautious cast of mind I am proposing is reasonable, is not costly, and ought to be politic.

I do not want to deny that there are cases where there may be great harm where we can do little. One of the current mysteries of biological science is the cause of the sudden and incredible decline of frog populations around the world. Frogs have survived for over 200 million years and so are a hardy species, but, herpetologists say, "frogs are the ideal creature to reflect the health of the environment . . . moving over their life cycles from water to land, from plant-eater to insect-eater, covered only by a permeable skin that offers little shield from the outside world."[1]

The concern is that their decline is "telling us about the environment's overall health,"[2] and because no one knows whether the cause is something toxic in the environment or is part of the natural cycle of frog populations, further research is needed.

Although not denying the need for such research, James Vial of Oregon State University says, "These declines, in many cases, are so catastrophic and are occurring simultaneously in so many parts of the world that we cannot wait for monitoring studies before we express our concern. . . . Because if we wait for a prolonged period, there may not be anything left to study."[3] He is right that waiting may mean the irreparable loss of the frog population, but, unfortunately, without any indication of what could be done, there seems nothing to do but further research. This example indicates one of the conditions under which it is appropriate to wait for further research to decide what to do, namely, when we know of nothing that can be done now to mitigate any harmful effects and so can only express concern.

But when we know something that can be done that may mitigate potential harm, then it *can* be reasonable to act, especially when the magnitude of the potential harm is significant, the magnitude of any harm to current interests is not, and we are mitigating the effects of a present harm in any case. But to see that acting can be reasonable in such situations of doubt requires a fundamental shift in the way we tend to *think* about public policy decision making—a shift that makes that talk show caller's question reasonable and the response to it inappropriate.

Such conceptual shifts are among the most difficult to achieve but the most readily accepted once made: What before seemed impossible now seems obvious. We can all smile about the child who, seeing a rainbow, said, "Poor rainbow! It's getting all wet." We know that rainbows are not the sorts of things that can get wet, and we smile at the child's delightful misconception. The child would no doubt have found it impossible to think how a rainbow could *not* get wet. A sort of giant invisible umbrella? However natural it is for a child to take a rainbow as a substantial object, like a chair or a dog, we have no difficulty understanding that the child is making a mistake. I will have succeeded if what I am suggesting about how we ought to reason regarding environmental matters seems as obvious.

Changing the way we think about something, however, when the way we think about it seems natural and appropriate to us, can be a difficult enterprise. If you take it as given that when someone asks a question of the form, "Do you want to do such-and-such?," they mean to say that they want to do it and wonder if you do too, you will be nonplussed by someone who asks that question just to see what you want to do. When you say in response to such a question that you are willing to visit your spouse's parents, for instance, but say so with a tone of voice that indicates reluctance, you will thus be surprised if your spouse, sensing your reluctance, then asks, "Do you really want to go?" Since you hear queries for information about your preferences as polite statements of preference, you will think that your spouse first stated a preference to go and then, after you reluctantly agreed to go, stated a preference not to go. It will look to you as though your spouse is either thoroughly unclear about what he or she wants to do or has forced you to make clear your reluctance to visit his or her parents and done so without any other agenda, just to trip you up.[4] Much miscommunication seems to turn on the underlying suppositions that form the background for our utterances, and changing what seems a natural and appropriate reading of what someone says can be a difficult matter—especially when the underlying suppositions are difficult to see *as* suppositions but are taken to be part of the meaning of the utterances.

A major difficulty with displacing the mode of decision making dominant in our current natural social artifact of waste management is that it is part of that social artifact and as such is normative for us. It is part of what it is to be "scientific" and "reasonable" about such matters, and so we may find it difficult to see it as a contingency, as a feature of that artifact that could be changed even though it both informs the artifact and is sustained by it. It is also pervasive—not like the simple mistake of taking a rainbow for a substantial object, but part of the fabric of our whole way of conceiving of how to come to grips with environmental issues. It directs the way in which we gather evidence, the way in which we weigh it, the criteria for funding projects, the very form of dialogue that must take place in any public policy discussion, everything that affects public policy about the environment. It even determines whom we pay attention to, as the talk show speaker made clear in discounting the caller's suggestion and so discounting the caller.

We can see the contrasting modes of thinking in different reactions to recent evidence that soot is responsible for up to 60,000 deaths a year in the United States. The federal government's standard "does not consider air hazardous until it reaches 150 micrograms of such particles per cubic meter," but recent studies have found a correlation between soot particles and the incidence of death at a significantly lower number of particles. For instance, in Utah Valley "the death rate . . . rose in proportion to particles in the air, on a steady slope in which each 100 micrograms of particle pollution increased the death rate 16 percent," and the pollution averaged "47 micrograms, less than a third of the Federal standard." The "particles were directly linked to health effects and the increasing or decreasing of ozone or sulfur dioxide were not." Similar studies showing a similar correlation were undertaken in Steubenville and St. Louis, and a study in Philadelphia showed a 7-percent increase for each 100 micrograms of particles.[5]

What was discovered was a correlation, and in response to that, Dr. Jonathan Samet of the University of New Mexico pointed out, "What is missing is understanding the biological basis of the effect they are describing." He then went on to say, as we saw, "We shouldn't make major policy decisions about causes of death based on the kind of evidence we have so far." Why not? Dr. Samet said that "he believed that there still might be other ways of explaining the deaths linked to pollution—for example, combinations of factors like respiratory illness, coming with a heat wave, at the same time as a pollution episode."[6]

The cast of mind I am suggesting he would no doubt consider "unscientific" because I am urging that we ought to make "major policy decisions" on the basis of the evidence so far available: Such particles *may* be the cause of many deaths. It is true both that we do not yet understand the biological basis of the correlation uncovered, if any, and that there may be other ways to explain the correlation. We would be in a better position if we knew that there was more than a correlation, that the particles were the *cause* of death or such a significant factor in increased mortality that decreasing the incidence of particles would have a beneficial impact on mortality rates. What we do know is the following:

1. "Fine particles are thought to pose a particularly great risk to health because they are more likely to be toxic than larger particles and can be breathed more deeply into the lungs."[7]

2. The correlation between mortality and "the levels of inhalable, fine, and sulfate particles" was "robust" in a variety of different settings.[8]

3. Although "it is possible that the observed association was due to confounding—that is, that it resulted from a risk factor that was correlated with both exposure and mortality . . . the association of air pollution with mortality was observed even after we [the study group] directly controlled for individual differences in other risk factors, including age, sex, cigarette smoking, education level, body-mass index, and occupational exposure."[9]

However, we do not know *why* this is so. We are not even completely sure why fine particles are a greater risk to health than less fine particles. We only know that they are *likely* to be. We do not know what the causal explanation is for these correlations, and so we cannot know for sure that there is a causal explanation. So, for all we can know, the observed relation between increased mortality and the incidence of such fine particles may be nothing more than an artifact of some other causal factor that those who did the study failed to find.

What Dr. Samet is suggesting is that nothing be done until further research determines if there is a biological basis for the correlation. He thinks it a *mistake* to base any "major policy decisions about cause of death on the kind of evidence we have so far." I am arguing that lack of the sort of knowledge he demands should *in itself* not preclude our acting to mitigate the harm that the studies so far done show may be occurring.

We should be clear about the nature of this disagreement, for otherwise we may find ourselves trapped in a thoroughly confusing misunderstanding, like the couple originally engaged in deciding whether to visit the in-laws. From Dr. Samet's perspective, from within, that is, the model of decision making he must think appropriate for major public policy decisions, what I am suggesting must seem both unreasonable and immoral. It must be unreasonable from his perspective because I am suggesting that we act without full knowledge, and it must be immoral because I am implying that we might act to change the existing situation and so cause harm to existing interests without knowing that harm is actually occurring *because* of the existing situation. Behind these two judgments from that perspective lies an immense set of arguments to support such claims as that it is morally wrong to cause harm without a showing that one's action will prevent or mitigate harm, that acting without knowing that one's act is responsive to a real problem is unintelligent because, for instance, we cannot know that we are not increasing our risks rather than diminishing them, and so on. Dr. Samet has at his disposal all the arguments that justify the scientific method as well as the compelling thought that we ought to act with knowledge and forethought rather than on whim, hope, or bias, prejudice, *especially* in regard to major public policy matters when mistakes will not be limited but are likely to harm large numbers of individuals, corporations, and the body politic.

But from the perspective I am arguing for, Dr. Samet's objections to acting now seem themselves unreasonable and even immoral. It is prima facie unreasonable, I would argue, not to act to mitigate possible harm, and it becomes more and more unreasonable the greater the potential harm, the more likely it is to occur, the less we must do and the less harm we must cause to mitigate the potential harm, and so on. This is a position we all adopt all the time—whenever we drive, whenever we traverse a slippery flight of stairs and tailor our steps to the potentialities, and so on. And it is morally wrong, I would argue, not to act to mitigate possible harm when the magnitude of the possible harm is great, the cost to present interests is small, and so on.

But to engage on this level of claim and counterclaim about which position is reasonable and which position is moral is to misunderstand the nature of the disagreement. What is in dispute is the sort of model of decision making we should adopt for public policy decisions about the environment. To attack one another from *within* our respective models is to get caught up in a fruitless encounter, such as that between the husband and wife. I do not mean to suggest that someone like Dr. Samet is unreasonable from *within* the mode of decision making he presumes appropriate. So what is at issue between us cannot be well captured by either of us saying of the other, "You are unreasonable," or "Your recommendation is immoral."

The conflict will be resolved only by our seeing that competing modes of decision making are at issue and that one model should not be privileged merely because it seems natural and compelling. We should not get trapped within the one model, acting like that Englishman I have mentioned, who, after going around the world, was so glad to be back where people spoke the way they thought. The model I am suggesting ought to drive public policy decisions about the environment is not an oddity, but one we use all the time—in driving when we slow down behind a somewhat erratic driver, in planting a bulb in one place rather than another, and so on. For in such situations we always have what the dominant mode of decision-making would consider an epistemic shortfall, and yet we can make perfectly reasonable decisions—modulating our judgments as the evidence and lack of evidence require, as the risks increase or diminish, as the potential harms increase or decrease in number or magnitude.

I have suggested that the conflict between these two models is in some ways like that between essentially contestable concepts, essentially unresolvable, yet with each "sustained by perfectly respectable arguments and evidence."[10] We would then be unable to choose between them in any rational way, and each would vie for success, each would sometimes win, and the success of each would be looked upon by advocates of the other as one more example of the machinery of decision making being captured by some unreasonable interest group—making decisions not just on the basis of "philosophical predilections"[11] but against common sense.[12]

Such a conflict of conceptions about how to make decisions might explain much that has been happening over the past several decades—clear examples of decisions for which more research is needed before anything can be done, clear examples, as with CFCs, that the risk of waiting for more research is too high to justify the wait, and, as we might expect when there is competition between essentially contested concepts with no decision-procedure to choose between them, a great deal of confusion about what the basis for such decisions ought to be, marked by frustration and name-calling on the part of whoever happens to be on the other side of the issue.[13]

Indeed, if this clash between competing conceptions is like that between essentially contested concepts, we can expect it to continue, unabated, as we continue to come to grips with environmental problems. The main four laws "that help form the foundation of . . . environmental policy are up for renewal in the next two years," the "Clean Air Act, the Endangered Species Act, the Superfund law for cleaning up toxic wastes and the Resource Conservation and Recovery Act for disposing of hazardous chemicals." One of the major questions before the Senate Committee on Environment and Public Works is a version of the question we are addressing here: "What should lawmakers and regulators do after they discover a new environmental problem that seems dangerous but before careful scientific analysis has shown the actual degree of risk?" And it should occasion no surprise that the sides are already in place, each apparently taking up one of the conceptions of how to make decisions. On the one hand, Senator Daniel Patrick Moynihan "has introduced a bill that would require the Government to amass much stronger scientific proof and convene expert panels of scientists to consider a new environmental rule before it is issued." On the other hand, Dr. Adam Finkel of Resources for the Future, says, "As a scientist, we always hope to have more research to answer complex questions. But in 1993 we don't have the scientific basis for rejecting the current approach, which says we should be prudent when faced with uncertainty."[14]

But although the dispute between these competing conceptions of how to make decisions in the face of doubt has some of the markings of a dispute between essentially contested concepts, not resolvable by rational means, I think we have some bases for choosing when to use which model of decision making. We have, in short, a decision-procedure for choosing which decision-procedure to adopt regarding environmental matters. I have made use of it in suggesting in regard to whatever is happening to frogs that our best course of action is more research, in arguing that our actions regarding CFCs were correct, in arguing that we should do what we can to minimize the use of toxic heavy metals. When we have *no* way of knowing what to do, rather obviously, we need to know more if we are to do anything at all that may address the problem, but when we have some evidence of a correlation, we at least have *that* much knowledge, and when the magnitude of the potential harm is great and the cost of change relatively small, that amount of knowledge *can* be a sufficient basis for action—especially when the correlation is "robust."

We can look at this way of proceeding in a more general way, borrowing from engineering the concept of an error-provocative design.[15] A design provokes errors when it is so counterintuitive that even the most intelligent, well-trained, and well-motivated person is likely to make a mistake. So a knob for a stove burner that turns to the left rather than to the right provokes errors because the vast majority of people presume, and have become accustomed to having that presumption borne out, that we turn things on by turning them to the right.[16] The ideal would be a design in which no operator errors were ever possible, so well-designed that even the least intelligent, the most poorly trained, and the least motivated would not err. This is not an ideal we are ever likely to reach, but we can so design things that we can reduce the incidence of features that provoke errors. If we examine the natural social artifact in which is embedded the decision-procedure that requires that we have all relevant information before we "make major policy decisions," we shall find almost nothing, no design feature, we may say, that prevents even the most intelligent, well-intentioned, and well-informed person from doing something that will later cause untold harm to others. We can apply this concept of an error-provocative design both to the decision-procedure we must use in this society to come to grips with our environmental problems and to the very way we tend to think about such issues.

First, we might think our political process a design that provokes, from the range of all possible solutions to a problem, those solutions which best maximize the chances for re-election (decisions after the next election, if at all; decisions after "a proper study" has been made; decisions in which the bad effects are far removed and the good effects occur before the next election; and so on).[17] Because there is no guarantee that the solutions that the political process provokes will be best for the problems—

indeed, because such solutions will probably not be best for environmental problems, which tend to require long-term solutions with the benefits far in the future and the costs up front—we may think of our political process as an error-provocative design. It has virtues that would be lost were we to have a redesign, as an engineer might put it, to mitigate its worst error-provocative features, but its virtues should not obscure for us its faults.

Second, the very way we tend to think about how to respond to environmental problems may be viewed as a complex natural social artifact that provokes certain sorts of errors. It has virtues, of course. It drives essential research, for instance, and so it must form an essential part of any stance toward environmental issues—to drive research to find the causes of the frogs' demise, for instance, to find the basis of the correlation between soot and mortality rates, if there is one, and so on. But, as I have argued, the dominant mode of thinking about how to respond to environmental problems also has at least two unfortunate effects. (1.) Demanding all relevant information before we act guarantees that if there is a problem, the problem will get significantly worse before all the evidence is in that would justify action to mitigate any harm. The requirement that we *know* that asbestos causes significant harm to the health of those working with it effectively precluded any attempt to mitigate the potential harm until many years later, when a great deal of harm had been done, with huge costs to the health of many of those exposed to asbestos and with equally large costs to remove or immobilize the asbestos in the environment that would not have been there had action been taken sooner. The requirement that we *know* how CFCs deplete ozone before acting to diminish the amount of CFCs leaking into the atmosphere would mean that we would be unable to act for a decade or more, as we gathered the relevant information. By that time, any potential problem would have become significantly worse, with the additional ozone depletion during that time causing more cancer, for instance, and with the additional CFCs in the atmosphere taking that much longer to dissipate, causing that many more problems over that much longer time.[18]

(2.) Demanding all relevant information before acting also entrenches existing interests, making them immune to change barring a showing of significant harm. The model thus effectively works the way a precedent works in the law. As I have argued, a court's deciding one way or another justifies reliance on that decision by those affected by it so that overturning the decision requires showing not just that it was wrong, but that it was wrong enough to justify the disruption to the expectations and actions it engendered. Similarly, what drives decisions about what to manufacture, what capital investments need to be made, and so on is generally laissez-faire economics with a due regard in our society to the possibility of lawsuits. A demand for all relevant information before acting both encourages such laissez-faire decisions,[19] because what would be required to prevent someone's deciding to do whatever he or she wants would be *knowledge* that the act produces harm, and it entrenches the interests created by those decisions. For requiring that we have knowledge of real harm before we act effectively produces a high hurdle that must be overcome to mitigate any harm that may be occurring because of existing interests.

Existing interests not only have priority simply because they are existing so that we need a *showing* of a significantly greater harm than would be produced by harming the existing interests, but also our model of decision making makes the standard for a showing of harm higher. It is not enough to show that existing interests risk harm to health, for instance. What counts as "a showing of significantly greater harm" is *knowledge* of, for instance, how it is that soot causes increased mortality.

Thus, the very way we think about how to respond to environmental problems and the system of decision making we must use to respond to those problems provokes a great deal of harm—by delaying a response to potential harm, by favoring solutions with short-term benefits even if there are serious long-term harms, and so on. It is the *design* of this way of thinking about and responding to environmental problems that causes these problems, and a corollary of having a flawed design is that one must thus rely on the intelligence and good will of everyone to become informed enough to make it work without serious harms.

Unfortunately, many environmental problems are caused by what, in retrospect, can only be considered a great deal of stupidity, such as dumping steel canisters with toxic metals into the ocean off California[20] or by siting a tanker terminal in Prince William Sound and not taking any precautions for a possible disaster. Many are caused by a lack of information or a failure of good will. DuPont examined polybrominated biphenyl (PBB) as "a base for flame-resistant synthetic fibers in such things as children's sleeping garments" but discovered that it "caused liver enlargement in rats" and that it "remained in body fat and would probably be biomagnified."[21] The company gave up on the idea, judging it not worth it to investigate the harm PBB might cause humans. Meanwhile, the Michigan Chemical Company proceeded to manufacture PBB as a fire retardant for such things as micro-fiche readers and to "discharge PBB into the Pine River."[22] Given the requirement for all relevant information before it can act, the relevant Michigan governmental agency, the Michigan Department of Natural Resources, could not prohibit the discharge without a *showing* that PBB is harmful to humans. When our way of approaching environmental problems makes us heavily dependent on the good will, and, I want to add, common sense, of various companies to do what they can to mitigate harm, we invite harm: Our approach provokes it.

But the worst problem with demanding all relevant information before we "make major policy decisions," while we continue to make public policy without all relevant information merely by continuing to act, is that it is incoherent. We cannot remove the sting of that incoherence by prohibiting all action until all relevant information is available, for, as I have argued, we must continue to act regardless, doing something with our waste, for example, even as we wait to determine what we ought to do with it and so effectively making policy that will have a deadweight of its own when we do come to decide what we ought to do. It is equally incoherent to opt for the model of decision making I am suggesting to respond to a problem, acting in the face of some doubt about its causes, and yet allow for interests to continue to entrench themselves willy-nilly, without a due regard for the possible harms that may result.

What is coherent is to adopt the cast of mind I am suggesting for our general way of moving through this world both for the capital investments we must continue to make in factories and other various forms of development and for the environmental problems that will inevitably arise. The proper cast of mind requires acting now to mitigate future harm that may occur.

It is this cast of mind that, I suggest, lies behind recent efforts to identify ahead of time what will become environmental problems and work out some compromise now to ensure that the problems will not become serious.[23] All that is needed to act now to find ways of preventing potential harm is some evidence of potential harm—the greater the magnitude of the potential harm, the more reasonable it is to reach an agreement before permitting development, and, for those problems that will inevitably arise, no matter how cautious we are, the greater the potential harm, and the less expensive the remedy, the more reasonable it is to act. We can thus mitigate the long-term effects of any harmful decisions we may make because we need only good evidence of possible harm before we are justified in acting, both to prevent some potential harm from becoming the object of an existing interest and to respond to potential harms.

We need to adopt this cast of mind self-consciously, however, fully aware of why we are doing it. Otherwise, among other problems, we shall find ourselves without the clarity of purpose and of vision that will give unity to our actions and that ought to permeate everything we do regarding the environment. We may otherwise find ourselves caught up short by someone playing the epistemological card in a situation of doubt, like that regarding soot, but using it to block a timely response rather than to ensure intelligent action even in the face of doubt.[24] The card may seem perfectly reasonable and even appropriate, and so we may miss that the deeper significance of the play is to forestall "major policy decisions." If we are not thinking clearly about *how* we ought to think, we may end up confused about what we ought to do.

But clarity about what we are about is of prime importance, especially if we are in transition between two ways of thinking about environmental matters and when we face enormous problems created by our past environmental practices and must try to work out how best to proceed given the expectations created by those practices and the real harm to existing interests, especially economic, that will be caused by continued change. Cases in point are the attempts to "help restore the endangered Everglades ecosystem,"[25] to continue the enormous progress that has been made in cleansing the air in Los Angeles,[26] or to restore San Francisco Bay and the Sacramento Delta and thus to help ensure the quality of the drinking water of 20 million people and the survival of a large variety of fish by diverting large amounts of water from farms and cities in Southern California that now draw on the water that supplies the Delta and the Bay.[27]

Such attempts to come to grips with past practices, with continued development, and with the environmental problems that will occur regardless of the best of plans can be done well or poorly. Those most likely to succeed are those which take advantage of mutual interests, building on what competing interests have in common[28] and aiming for that politician's dream of satisfying everyone concerned[29]—working out the calculus of interests and rights in such a way that all recognize that the public good is served and that each has contributed fairly to the result.

But even with clarity about how we ought to think about environmental problems, we will not be free of problems, for there are difficulties inherent in the cast of mind I am arguing for. Consider again the problem of soot and its correlation with higher rates of mortality. To question Dr. Samet's conclusion about what is needed "to make major policy decisions" is to raise the issue of how much evidence of possible harm is enough. That is not an easy issue to respond to if, as I claim counter to Dr. Samet, we can act reasonably and ought to act in some ignorance, for then we must determine how much ignorance is acceptable. Is having a correlation enough, or is more required? Must the correlation be robust? And how much of a correlation makes it a robust one?

Finding the right degree of caution about the right objects at the right time and in the right way is a difficult matter.[30] We may err in a variety of ways, taking something to be dangerous when it is not, taking something to be more dangerous than it is, acting too late to mitigate harm, being overly cautious, or taking what turn out to be undue risks, attempting to mitigate the harm in the wrong way,[31] and so on. Responding with considered judgment to complex issues is thus a delicate as well as a difficult matter, one in which we may readily make mistakes.[32] The issue is complex. Should we err on the side of caution or not; what values—human health, economic interests, or what—ought we to consider when making that decision; and how are we to weigh them one against another?

That mistakes can be and have been made in such situations is no argument that we ought to opt for an incoherent model of how we ought to make decisions that carries the risk of worse harms than occur from any of the mistakes that come from exercising our considered judgment to avoid risk to human health.[33] As it turns out, many cases are straightforward and clear.

In the case of soot, for instance, other factors conspire to a resolution of what we ought to do—without requiring more evidence. Unlike the case of the disappearance of the frogs, this case provides us with a likely suspect—soot particles. But having a likely suspect would not be enough to justify doing anything if there were nothing that could be done, or even if doing something were so unbelievably expensive that it would, to take a worst-case scenario, bankrupt the companies whose factories are responsible for releasing the soot. We would then have to weigh the benefits of change against the loss of jobs, of future opportunities in the industries affected, and so on. Knowing there is a correlation is not enough, in short, to justify making major public policy decisions.

But this case has two additional features that ought to help us come to a public policy decision. In the first place, no one is suggesting that the particles themselves are *good* for anyone. Quite the contrary. Having a standard in which some percentage of particles in the air constitutes a hazard is evidence that they are indeed hazardous, although we may disagree about the magnitude of the hazard and about how many particles constitute a hazard. Even with that disagreement, the presumption is that fewer would be better than more, and if the finer particles are more likely to be hazardous, and the current regulations are biased in favor of screening out larger particles, then that is some argument for changing the standards—at the least changing the bias to exclude more of the finer particles.

In the second place, it is relatively inexpensive to remove the particles from the air. In response to the evidence, Dr. Joel Schwartz, an epidemiologist at the EPA, said, "If we consider how much protection we can get for our money, we can do a lot with a relatively small investment." He went on to say "that taking particles out of smokestack emissions was relatively cheap, and some newer devices such as 'bag houses,' filter-lined bags like vacuum cleaner bags, now can draw as much as 99 percent of particles out of the air without companies having to expend great sums on catalytic converters."[34]

So, although the mode of thought appropriate for environmental decisions is complex and so open, as any decision-procedure is, to mistakes that it takes care to avoid, it is coherent and does not provoke errors of the sort that mark the decision-procedure it displaces. For the error would rather be *not* to act on the basis of the kind of evidence we have so far regarding the relation between soot and mortality, especially as we know what we can do and as what we can do is relatively inexpensive and is a good thing in any event. If it is better to have fewer micrograms of such particles in the air than more, and if we can get fewer and harm our current interests little, then it is perfectly reasonable to act to cut the number of particles in the air and immoral not to, given, again, how inexpensive it is to protect ourselves against the possible harm and how, in acting to protect ourselves against that, we also act to mitigate any hazard we face in having such particles at all.[35] More research needs to be done, but making public policy should not be a hostage to the continued gathering of data—especially given, as in this case, the configuration of such conditions as the inexpensiveness of a remedy. Waiting privileges the status quo and ensures that if there is a causal relation, more will die needlessly.

The form of reasoning I think appropriate for such cases as that of soot drives equally reasonable decisions about such environmental matters as whether to site an oil tanker terminal on Prince William Sound or reduce the amount of heavy metals in the waste stream. When we turn this cast of mind to our natural social artifact of waste management, for example, and to features within this artifact, we see that although it supports a different set of alternatives from the mode of thought it is meant to replace, what it implies is not at all radical but quite sensible.

If we look at all our waste, for instance, and ask, "But what are we to do with all this stuff?," we get the wrong answers. We should look at the natural social artifact that constitutes the way we use the resources of the world and handle the waste that we produce. We should look at the whole process from our creation of waste to our disposal of it. For we are waste producers on a massive scale who put at the very least our own health at great risk by our practices. What we need to ask is not what we should do with all the stuff we produce but how we can restructure our natural social artifact, and thus our lives, to produce less and manage better what we do produce.

Making clear the proper object of our concern tells us what are our real alternatives and provides us with guidance about how to rank the alternative ways we have of handling waste—landfilling, reducing, recycling, dumping, and burning. Our ways of handling waste are means to different ends and to different ways of moving through this world, and if our long-term end is to restructure our natural social artifact of production and waste managment so that we have less of a problem with waste in the future, then we must pick the proper means to achieve that end. The cast of mind I am urging is a means to that end. What is required is that we attempt to *think* in a way different from that in which we have been urged to think about environmental issues and, with that new cast of mind, approach our world differently—and change it for the better.

NOTES

Preface

1. *The Ideological Origins of the American Revolution* (Cambridge, Mass.: The Belknap Press, 1982), pp. 144–59.
2. *Solid and Hazardous Waste Incineration: An Analysis for Citizens and Policymakers* (March 1987), unpublished.

1. Making the Best Choices

1. The Department of Agriculture first used 5 parts per million (ppm) as its basis for banning milk, then moved to the 1 ppm in milk fat "mandated by the U.S. Food and Drug Administration," and then, in November 1974, to 0.3 ppm. The last measurement was determined in part by what were claimed to be the limits of instrumentation—"0.3 ppm . . . with reasonable accuracy." The effect was that "the allowable level of poison was established partly by the sophistication of the equipment rather than by any real knowledge of its toxicity"—an unfortunate but inevitable consequence if we really could only measure to that level of adulteration ("Cheap Chemicals and Dumb Luck," *Audubon*, Vol. 78, No. 1 (January 1976), p. 115). Others claimed that measurements were possible to 1 part per billion (ibid., p. 116).
2. What looked to be a scientific disagreement over what amounts of PBB were measurable had an effect on the ways in which the differing parties responded to the problem. A more stringent limit would have banned far more foodstuffs than the State's. What appeared to be a scientific dispute had significant economic and political implications.
3. The Michigan Department of Public Health claimed that "there was no immediate identifiable human [health] problem" ("Cheap Chemicals," p. 115). But it did not need to make this claim to argue that without knowledge of a relationship between any "identifiable human [health] problem" and PBB, it would not be rational to cut even more PBB-laced foodstuffs from the food chain. The De-

partment could have conceded, that is, that health problems were *associated* with PBB. All it needed to deny was that PBB was known to be the cause of those problems. It instead went so far as to say, without evidence, that "there was no danger to persons who ate or drank these products because the contaminated milk was diluted with pure milk" ("A Poison Quarantines Cows and Hens," *New York Times* [June 10, 1974], p. A45). An additional objection is that rather than find the sources of contamination and remove them, the state was willing to allow the PBB to be diluted—a usual way of handling toxic pollution, dissipating it by spreading it thin far and wide rather than concentrating it and then disposing of it.

4. For a more thorough but still brief analysis of Michigan's problems with PBB, see my "Philosophy and Public Policy: A Look at PBB," *The Michigan Connection* (Newsletter for the Michigan Council for the Humanities), Spring 1977, pp. 7 and 12. I examine the case in somewhat more detail in Chapter 6.

5. It is an essential part of this vision that only those propositions which have gone through this winnowing process by the scientific community, and so are reproducible by any competent scientist in the field, are to be allowed to count as premises in making rational decisions regarding those public policy issues for which such propositions are relevant.

6. According to Dr. Jonathan Samet of the University of Mexico, quoted by Philip J. Hilts, "Studies Say Soot Kills up to 60,000 in U.S. Each Year," *New York Times* (July 19, 1993), p. A16. We examine this case in detail in Chapter 7.

7. "An effective way of thwarting practical decisions is to insist on the highest standards of scientific rigor" (Baruch Fischhoff, Paul Slovic, and Sarah Lichtenstein, "'The Public' vs. 'The Experts': Perceived vs. Actual Disagreements about Risks of Nuclear Power," in Vincent T. Covello, W. Gary Flamm, Joseph V. Rodricks, and Robert G. Tardiff, eds., *The Analysis of Actual Versus Perceived Risks* (New York: Plenum, 1983), p. 243).

8. Researchers at the Department of Agriculture have recently discovered that an AIDS-like virus—BIV, or bovine immunodeficiency-like virus—is relatively prevalent in cattle in the United States. There is a concern about its effects on human health, on the $68 million-a-year semen industry, and on the health of those cattle not infected. Little research has been done on any of these issues or on the possible cause or causes. One suggestion is that the heavy use of such antibacterial agents as antibiotics on cattle—at the rate of $300 to $500 million a year—may be causing "immune suppression in adult cattle," but this has not been studied. What is known is, for instance, that in the light of possible infection by a disease whose effects are unknown, European importers of semen from American cattle may cut back on their imports (Keith Schneider, "AIDS-like Cow Virus Found at Unexpectedly High Rate," *New York Times* [June 1, 1991], p. A8).

So the question concerns what we ought to do, given what we know and what we do not know. The presumption that we need full information to act would power the research necessary to answer such questions about the cause or causes of BIV and also justify not doing anything until the evidence is in. We do know that like AIDS, BIV is spread through the blood. So one way to inhibit its spread among cattle is to ensure that blood is not shared by banning the use of the same needle for drawing blood or for successive injections of drugs. That would diminish the likelihood of BIV's spreading. But because we do not know that BIV is causing any harm, can only know, given the current state of knowledge, that it may cause such harm as causing the exports of cattle semen to drop, the presumption that we need full information to act prevents us from even taking such relatively minor cautions.

One may object that such a ban would be difficult to enforce or that, in the light of no obvious harm, it would be difficult to educate those responsible for drawing blood and injecting drugs not to use the same needle, but these are objections about the difficulties of doing something that is otherwise "a good thing." The presumption that we need full information to act means that such a ban is not a good thing.

Coming to grips with the problem of how rational it is to await full information before acting takes on some importance when humans may be at risk. Many domestic cats are infected with an AIDS-like virus, called FIV, or feline immunodeficiency virus, and, as one veterinarian who has an infected cat said, "There is no evidence that cats can transmit FIV to humans . . . but I always hedge on that when you have a baby. To me, it's an unhygienic situation" (Fox Butterfield, "AIDS-Like Virus Found in Many Domestic Cats," *New York Times* [July 13, 1991], p. A7).

9. Paul Brodeur, *Outrageous Misconduct: The Asbestos Industry on Trial* (New York: Pantheon Books, 1985), p. 30. See pp. 10–14 for a brief statement of the state of knowledge about the dangers of working with asbestos in the early 1900s.

The argument of Brodeur's book is that the asbestos industry acted outrageously in claiming that it only came, and only could have come, to see the danger in working with asbestos in 1964—much too late, by its lights, to hold it liable for the harm to workers exposed before that time, especially when its products only contained 15 or so percent of asbestos. The industry could, and did, defend itself with two claims about its position in the 1930s: (1) it did not know that exposure to asbestos would have long-term harmful effects; and (2) it did not know that exposure to such a small percentage of asbestos would be harmful because there was no information available about the effects of such limited exposure. But underlying these claims of the state of their knowledge is a third assumption the industry was making: not enough information was available to be sure that we should ban the use of a beneficial product because of harm it *might* cause. The asbestos industry was making an assumption about how to make decisions in the face of doubt.

Underlying Brodeur's analysis, giving it the power it has, is a presumption that such knowledge as the industry did have in the 1930s, or could have had, was sufficient to justify, among other things, warnings to workers. Of course, that manufacturers of asbestos ought to have acted on that assumption is effectively what courts decided in the series of cases Brodeur examines. In short, it is not only how much knowledge the asbestos industry had that is at issue, but what they ought to have done with what knowledge was available. They presumed a stance about how we ought to make decisions when in doubt about harms—a stance I shall argue is mistaken, in this case and others.

10. These and subsequent remarks may suggest that the issue regarding the dangers of asbestos are clear and that there is no disagreement in the scientific community about the matter. But that is far from the case. See Joseph Hooper, "The Asbestos Mess," *New York Times Magazine* (November 25, 1990), pp. 38ff.

11. Recent studies of the health records of over 35,000 workers at a government bomb plant in Washington indicate that even small doses of radiation pose a health hazard. There are three claims: that "even small doses of radiation are four to eight times more likely to cause cancer than previously believed," that "people are far more vulnerable to radiation-induced cancer if the exposure comes later in life," and that "radiation . . . in small doses over time may carry a higher risk of cancer than radiation delivered in a single dose" (Matthew L. Wald, "Pioneer in Radiation Sees Risk Even in Small Doses," *New York Times* [December 8, 1992], p. A14). These claims directly contradict earlier assumptions that have marked how we have protected workers' safety ever since the nuclear age began. If these claims are correct, and it cannot be certain that they are, at least in part because they are based on evidence of health records that the government has only recently released and that consequently have not been well studied by researchers, then the price of those earlier assumptions is that there will be an excess of over 200 deaths attributable to cancer at that plant (ibid.).

12. The Environmental Protection Agency (EPA) has recently asked DuPont to put off phasing out its production of CFC-12, a refrigerant for air conditioners, from 1994 until December 1995. The aim is to ease the transition from the old to the new coolant. Because the new air conditioning fluid is not compatible with old air conditioning units, those car owners whose air conditioners fail will be faced with the choice of driving without an air conditioner, purchasing a new air conditioner for "as much as $1,000," or buying a new car. The EPA's aim in asking DuPont to continue production of what it has decided should not be produced is clearly to minimize the disruptions that would be caused by cutting off production by the earlier date (Julie Edelson Halpert, "Scarcity of Car Coolant Could Prove Costly," *New York Times* [December 26, 1993], p. F5).

13. For an excellent discussion of the particular problems and interests of a bureaucracy, see Terry M. Moe, "The Politics of Bureaucratic Structure," in *Can the Government Govern?*, John E. Chubb and Paul E. Peterson, eds. (Washington: The Brookings Institution, 1989), pp. 267–329.

14. Whatever that may mean—or whatever the official can get away with making it mean. The concept of the public good, like so many concepts in politics, is best viewed as essentially contestable. People disagree about what such concepts mean, both theoretically and practically, but such disagreements are not psychological, or verbal, or "philosophical" but "perfectly genuine." These are disputes "which, although not resolvable by argument of any kind, are nevertheless sustained by perfectly respectable arguments and evidence" (E. Gallie, "Essentially Contested Concepts," *Proceedings of the Aristotelian Society*, Vol. LVI, New Series [1955–56], pp. 167–98; see also Ronald Dworkin, *Taking Rights Seriously* [Cambridge: Harvard University Press, 1978], p. 103).

Arguments about the concept of equal protection in the 14th Amendment have this character. In the terminology of John Rawls, the concept gives rise to competing conceptions (*A Theory of Justice* [Cambridge: Harvard University Press, 1971], p. 5). For instance, in regard to funding for grade school education, it is arguable both that it is unfair to deny educational opportunities to any child because of place of residence and that it is unfair to disadvantage those children whose parents live in richer suburbs and may have moved to those suburbs just to obtain better schooling for their children. The former conception requires the same amount of aid to each child; the latter permits inequalities founded on differences in property taxes provided that a basic minimum is met. These competing conceptions of what constitutes equal protection in funding public education are essentially contestable—each supported "by perfectly respectable arguments and evidence" but true competitors with no clarity about how to resolve the competition by rational discourse. An elected official who maintains that a decision was for the public good must not only show that, indeed, the decision was for *some* conception of the public good, but also that the particular conception is "the right one"—or at least one that will not be significantly contested by those who matter to the official, namely, his or her constituents.

One may look at this book as a sustained argument that we need to refashion our understanding of the public good to include a new way of making decisions about environmental issues. But, of course, my conception is *initially* as contestable as is the one I mean it to displace, and the argument's aim is by the end to make it so plausible and obvious that it will seem as natural and compelling as, for instance, that natural presumption that all relevant information is necessary for a fully rational decision.

15. I make use of this concept of an error-provocative design in greater detail in the concluding chapter. It is a helpful concept to apply to various decision procedures. A parent trying to allocate resources among his or her children would be using such a design if some one child always made the choice about which TV show all shall watch or about who gets the biggest piece of cake. Such a decision procedure is guaranteed to provoke all sorts of problems, not the least of which is that it will ensure that the wrong choice is often made, even if the child is well-intended and tries to do what is right. I owe my introduction to the concept to Dr. Jasper Shealy, III, Head of the Department of Industrial and Manufacturing Engineering at the Rochester Institute of Technology. He is not responsible for the use I make of it.

16. The principle is Aristotle's. He says that "a well-schooled man is one who searches for that degree of precision in each kind of study which the nature of the subject at hand admits" (Aristotle, *Nichomachean Ethics* [Indianapolis: Merrill, 1962], p. 5). This principle leads him to the conclusion that "when the subject and the basis of a discussion consist of matters that hold good only as a general rule, but not always, the conclusions reached must be of the same order" (ibid.). I draw the same conclusion.

17. The garbage barge was the *Mobro*, and it sailed fruitlessly for almost four months before returning to New York, load intact. Such odysseys are not as unusual as one might think. The *Mobro* "carried a mere 3,189 tons of garbage, but [a

ship called] the *Khian Sea* was loaded with nearly 15,000 tons of ash [and] wandered the seas for 26 months before it . . . dumped its trash" (William Bunch, "Where Will All the Garbage Go?," in *Rush to Burn: Solving America's Garbage Problem?* [Washington: Island Press, 1989], p. 76). These meanderings affect other forms of transportation as well. In April 1991 "a 32-car train carrying 2,400 tons of contaminated soil" left Freeland, Michigan, heading south to dump its load. It was refused at nine different dump sites in Ohio, West Virginia, Virginia, Tennessee, North Carolina, and South Carolina. Michigan dump sites had refused the soil that had been classified as hazardous because it had been soaked in acrylic acid in a spill from tank cars but had been reclassified as solid waste when it lost its capacity to ignite. While in South Carolina, the transport company was fined $21,975 for "letting some as yet unidentified liquid leak from one of the cars while it was there" ("Train Bearing Tainted Soil Rolls On," *New York Times* [April 24, 1991], p. A10).

18. The root idea is that we are concerned with hazards, and the core notions of a hazard are that of taking a risk (as in hazarding a chance) of taking a loss (as in a game of chance where the outcome is a loss of one's bet). So hazardous waste is material we throw away that puts at risk our health or physical well-being.

In a short discussion of the concept of a hazard, Christoph Honhenemser defines "hazards as a causal sequence of events which end with some kind of human or biological consequence, maybe death" (*The Analysis of Actual Versus Perceived Risks*, p. 324). I presume he means some causal biological consequence *harmful to our health*, but as his definition makes clear, defining hazardous waste is not an easy matter. For instance, is waste really hazardous if there is only the *slightest chance* of harm to our health—as the phrase "*some* kind of . . . consequence" may suggest? But we will not explore such problems here. It will be enough for our purposes to have clear cases of waste that are in fact hazardous if ingested—lead that enters our water supply, for instance.

19. Linda Greenhouse, "Justices to Decide if Ash Is a Hazardous Waste," *New York Times* (June 22, 1993), p. A19.

20. Ibid. The EPA has changed its mind about the status of such ash "so often," declared Chief Judge William J. Bauer, of the Federal Appeals Court in Chicago, "that it is no longer entitled to the deference normally accorded an agency's interpretation of the statute it administers" (ibid.).

21. Of course, the solid waste of a local landfill is, one hopes, not identical with the trash of one's home. The long lines of cars that form whenever a city has a special collection for hazardous household waste is a sign of that. For homeowners come in cars filled with the waste of basements and garages and attics that cannot properly be put on the curb. Stamford, Connecticut, had a "Household Hazardous Waste Collection Day" recently with 851 cars coming through, resulting in 259 drums being filled "with hazardous or toxic wastes," at a cost of $43,000 (George Judson, "A Cleaner Conscience, A Trash-Free Basement," *New York Times* [May 11, 1993], p. B5). I want the reader to keep in mind the trash of their household, including what they ought to dispose of through such a Collection Day—

even if they have not given any thought at all to the matter of what of their trash is hazardous.

22. The principle here is that obtaining quality trash is what will make recycling economically viable. Quality trash is trash in which the various components have not been mixed completely. As the manager of one recycling center puts it, "There is no practical way to process completely mixed materials. . . . Once something is mixed with trash, it becomes trash" (John Holusha, "In Solid Waste, It's the Breakdown That Counts," *New York Times* [March 31, 1991], p. F5).

Economic efficiency in recycling is more readily achievable if the waste comes already separated into its relevant parts, and the waste of factories can be more readily delivered in that way than the waste of homes. Indeed, factories may have an economic incentive to separate out the recoverable components of their waste. The long-term advantage is that recycling can come to be driven by economic considerations rather than by the push of a concerned public who may lose interest.

Manufacturers recycled twenty billion pounds of "the 38 billion pounds of [the] chemical wastes . . . produced in 1991" (Keith Schneider, "Manufacturers Recycling Half of Chemical Wastes," *New York Times* [May 26, 1993], p. A15). That figure of 38 billion pounds does not include "toxic wastes produced by the mining and oil-exploration industries, which the [Environmental Protection Agency] says generate the bulk of toxic chemical wastes" (ibid.). So there are two lessons here. The first is that what can be readily identified and kept separate, or separated, and so readily measured, has economic value because it replaces what would otherwise have to be purchased, because the cost of cleaning it up if it were dumped is prohibitive, or both. The second lesson is that although *some* industrial waste is easier to keep track of than the waste that enters the municipal stream, keeping track of the waste from some industries is a significant problem.

23. An international investigator, Eric Ellen, director of the International Maritime Bureau in London, said that one real "worry is that an irresponsible owner or a frightened crew, trapped in an enormously expensive search for a legitimate disposal site, may simply dump the material at sea" (Barry James, "Tramp Freighters Ply High Seas with Dangerous Cargoes," *International Herald-Tribune* [September 2, 1988], p. 1). The worry is driven by the realization that huge profits can be gained from making waste disappear. A West German firm, Weber Ltd., sold "1,500 tons of waste-laden sawdust for approximately $70 a ton" to a Turkish cement plant for its burners. Weber was paid between $450 and $510 a ton to dispose of the waste. Its own costs were about $110 a ton, and its clients in West Germany would have had to pay $560 a ton to dispose of it in West Germany. So everyone made significant profits except the cement plant, which ended up with tons of sawdust containing dangerous concentrations of PCBs ("The Global Poison Trade," *Newsweek* [November 7, 1988], p. 19). Dumping the sawdust at sea would have been almost as profitable—and caused less of a scandal.

24. In Stratford, Connecticut, the city used as landfill waste from Raymark Industries, which manufactured automotive brakes with asbestos. The "Raybestos

sludge" contains lead, asbestos, and PCB, and in one of the seven sites so far uncovered, the level of lead is "above 10,000 parts per million, 20 times the level considered safe" (George Judson, "What Were They Thinking Of? Yesterday's Clean Fill Is Today's Toxic Waste, That's What," *New York Times* [May 22, 1993], p. 21). As the headline of the article makes clear, the question current residents ask is "What were they thinking?," and the answer is that they were thinking "clean fill." Sites include a town park and "a middle school's playing fields," and $8 million have so far been committed "to cover and encapsulate the seven sites, reburying toxic materials that have worked their way to the surface over the years" (ibid.). The areas have been declared a public health threat, and a "public health advisory warns people to stay off the contaminated land and avoid eating seafood taken from Ferry Creek, a stream where Raymark Industries discharged waste for decades" ("U.S. Calls 15 Stratford Waste Sites Hazards," *New York Times* [May 28, 1993], p. B6). The city welcomed the announcement because it may mean Federal funding.

25. It is easy to find examples. The city of Haverstraw, along the Hudson north of New York City, arranged for U.S. Gypsum to dump scrap Sheetrock, or wallboard, in several places "to reclaim the carved-out earth" created by former clay mining for bricks. But when wallboard is exposed to groundwater, it decays and emits "hydrogen sulfide, which smells like rotten eggs in low doses and can be deadly in strong concentrations."

The problem that would create for the residents of the town would be bad enough, but the story does not end there. After the problem was discovered in the 1970s, one of the former sites was purchased and "re-zoned from industrial to residential use without an environmental impact statement." Houses were built on it, and now the owners are suffering the consequences. "On good days it gives off a nose-crinkling smell of rotten eggs; on bad days, residents say, the fumes are so thick they assault the eyes" (Lindsey Gurson, "18 Suburban Homes: A Dream Undermined," *New York Times* [May 13, 1993], pp. B1 and B10).

The home owners may have an even worse problem if a ruling by the Connecticut Supreme Court is used as a precedent for how to determine liability for such pollution. A woman inherited some land when her husband died on which "hazardous solid wastes had been dumped . . . by a Massachusetts trucking company that is now defunct." The Court held that she was liable "for abating the pollution on her land, the cost of which," the Court added, "may be in excess of the value of the land." That "draconian result" was mitigated slightly by a Connecticut law that would effectively limit her liability "to the fair-market value of the property" ("Court Holds Landowner Responsible in Pollution," *New York Times* [July 8, 1993], p. D2). If such a ruling were to be a precedent in other states, the effects on the buying and selling of property would be onerous.

26. Thus the City of Syracuse is faced with "decades of industrial dumping [in Onondaga Lake] that left a layer of toxic mire on the lake bottom, prompting the authorities to ban swimming in the 1940s and later to ban eating the lake's fish. Today the five-mile-long lake . . . is one of the nation's most polluted bodies of

water" ("Cost to Clean Poisoned Lake Staggers County," *New York Times* [May 6, 1993], p. B8).

Of course, the relevant authorities failed to act even when they knew they had a problem, and so the industries lining the lake continued and today to a large measure still continue to dump their waste—changing their practices only when they lose lawsuits that are expensive for any political authority to mount. The companies may claim they have some excuse because the city itself continues to dump both raw and treated sewage into the lake. It is a measure of the magnitude of the problem that a proposed solution would cost "$834 million over 20 years" and would help alleviate the problem caused to the lake only by diverting some sewage to the Seneca River (ibid.).

27. And it is not likely to do them much good in the short term and localized situations in which they now find themselves. The town of Thomaston in Connecticut is a case study in how past decisions can come back to haunt officials who may well have acted with the best of intentions. Among its other problems, it has become embroiled in "a Superfund lawsuit" against it and some other defendants for a total of $70 million. The suit "has cost $54,000 for lawyers so far," but, the city claims, it is subject to the suit only because it was following "the state's rules" regarding its landfill (George Judson, "Pollution Costs Torment a Town Already Reeling," *New York Times* [March 28, 1993], p. A34).

2. Natural Social Artifacts

1. The protest was pointedly against a particular corporation that had operated an incineration plant in Utah, among other places, and against the current state government of Pennsylvania for proposing that there be one in Allenwood. I have reproduced the capitalization of the originals, but not, by any means, provided them all: the streets are lined with so many of them that they are difficult to see as one drives by, and they are as various in design and ways of expressing the common sentiment as are, presumably, the inhabitants.

2. This is a main reason why I put to one side the many writings that call for us to undergo some sort of conversion and return to our "natural" relation with nature. I would be hesitant in any case about the fervor such calls for conversion require and produce, and nothing I shall argue depends on or requires any such conversion.

3. This is Lester W. Milbreath's conception in *Environmentalists: Vanguard for a New Society* (Albany: State University of New York Press, 1984). The last chapter is entitled "Can Modern-Day Prophets Redirect Society?," and in the introductory chapter, he says, "It would be helpful if today's society could find some modern-day prophets who understand, much better than ever before, how the world works physically and socially and who also have the breadth and depth of vision to develop a new ethical/normative belief structure that would enable humans to . . . live lives of reasonably high quality in a long-run sustainable relationship with nature" (pp. 6–7).

4. To suppose otherwise is to suppose "that the public is not rational" (Dr. Bailor, in the Panel Discussion, *The Analysis of Actual Versus Perceived Risks* [New York: Plenum, 1983], p. 332).

The presumption that the public is not rational regarding environmental matters seems a common one. Here are two examples chosen relatively at random from the anthology in which Dr. Bailor's remarks appear. In talking about the disaster at Three Mile Island, Anne D. Trunk and Edward V. Trunk write that "many were unable to utilize a logical thought process in arriving at a course of action" ("Impact of the Three Mile Island Accident as Perceived by Those Living in the Surrounding Community," ibid., p. 225), and in a more general analysis of how we respond to risk, Joanne Omang says that "people clearly are not rational about risk" ("Perception of Risk: A Journalist's Perspective," ibid., p. 168). Dr. Bailor goes on to suggest in his remarks that it is the premises that are at fault: "If the word rational applies to correct reasoning from specific starting points, I suspect . . . that the public may on the whole be entirely rational" (ibid., p. 332). The public reasons well, the suggestion is, but reasons from mistaken premises. If it had the right information, it would reason rightly. But I suggest we take as a working hypothesis not that the public is at fault here somehow, because citizens are either ignorant or unreasonable, but that we have competing visions of the appropriate mode of reasoning regarding environmental matters. It is an empirical question how "the public," whoever that is, reasons regarding such matters. It is not a question I investigate, but I think we ought to presume rationality rather than not and so direct any empirical studies to finding out exactly how it is that people do reason about such matters. My claim is that no matter how they do reason, they ought to reason in a way different from the way that causes us to have an incoherent public policy.

5. Omissions are causally as crucial as acts. Someone's not smoking is as causally relevant in creating a practice of smoking, a pattern of smoking and not smoking, as someone's smoking. I do not emphasize omissions in what follows because I do not think they play any more powerful a causal role than acts, but we should not forget that they play no less a role. In addition, I consider in Chapter 3 two different ways of omitting to act—deciding not to do something and not deciding whether or not to do something. Omissions are not all of a sort, that is, and there may be moral differences of some import in the way in which one does not act. Not smoking as a protest is surely different from not smoking because one does not want to.

6. I presume, without argument here, that natural social artifacts evolve and that they evolve in response to whatever pressures produce change. I will make this point again later, but I do not mean, in emphasizing what works and what is efficient, to prejudge what *general* evolutionary pressures produce change in all such artifacts, if any do. I mean only to give a sample of what kinds of pressures may do such evolutionary work. I do not even want to prejudge the issue of whether there are any general causal factors that hold for all natural social artifacts. That is an empirical question to be answered by looking at cases.

Looking at cases is of crucial importance for anyone concerned about changing a natural social artifact such as our present method of waste management. Understanding what causal factors are relevant and the ways in which they are relevant would give us a handle on how to initiate and sustain change. So my suggestions are meant only to mark out a certain way of looking at natural social artifacts: They are the product of acts and omissions, as artifacts, that is, and they are thus subject to change by changing the acts and omissions that produce them. What is necessary is to discover what motivates the various acts and omissions and what can encourage some motivations and discourage contrary ones.

7. How such a process is marked by intentional action is no easy question to answer. In this regard, see the detailed analysis of some of the problems in G. E. M. Anscombe, *Intention* (Oxford: Basil Blackwell, 1957). See especially the discussion beginning on p. 34 about the "man who moves his arm in pumping water into a house water-supply and is also doing other things with the pump handle at the same time" (p. vi).

8. I am tempted, regarding natural languages, to add "by the speakers of the language," but I leave to one side the question of who is involved in creating such a practice. I do not want to exclude the possibility that a natural social artifact that is a practice of a particular group of individuals may have been created, in part at least, by the actions and/or omissions of other individuals or groups. For instance, a stray foreigner may have, by mispronouncing an English word and having the mispronunciation picked up and carried on by native speakers, contributed to the language. I also do not mean to imply either that every speaker of the language necessarily contributes to it or that not every speaker does. Nothing I say about natural social artifacts hinges on how many individuals for whom they are controlling creates them or on whether they are formed by special groups, elites perhaps, within the groups for which they are controlling, and so on. These are intriguing questions, perhaps with helpful answers, for they raise the issue of whether we can fundamentally alter a particular natural social artifact by changing the behavior of a favored few—such as making it de rigueur to have higher hemlines by convincing those who are thought to set fashion to wear higher hemlines. But I do not press these questions here.

9. For a thorough analysis of how such change can occur, and of what its implications are for a particular practice, see Ronald Dworkin's analysis of how it is that our practice and our concept of courtesy may change. He distinguishes between what I would call the point of the practice and the relation of the practice to its point or points. He argues that "the practice of courtesy does not simply exist but has value, that it serves some interest or purpose or enforces some principle—in short, that it has some point—that can be stated independently of just describing the rules that make up the practice" and also argues that "the requirements of courtesy—the behavior it calls for or judgments it warrants—are not necessarily or exclusively what they have always been taken to be but are instead sensitive to its point, so that the strict rules must be understood or applied or extended or modified or qualified or limited by that point" (*Law's Empire* [Cambridge, Mass.: The Belknap Press, 1986], p. 47).

Dworkin goes on to examine how such a practice as that of courtesy can change over time in response to other changes. People may have thought that "the point of courtesy lies in the opportunity it provides to show respect to social superiors" and that "opinions may change about the nature or quality of respect," about whether respect can be shown classes rather than individuals, and so on (ibid., pp. 48–49). So the practice will change, evolving sometimes slowly, sometimes more quickly and more radically—as might happen were a revolution to occur and "a new order" be instituted in which no one is supposed to be anyone's social superior.

Dworkin's concern is with adjudication and thus with how judges may make decisions appealing to the same concept—equal protection, for instance—even when circumstances have changed and the decisions may indeed conflict with one another. A legal system is a particularly nice example of a natural social artifact whose very point of securing rules so that people can plan their futures mitigates against change (see Lon Fuller, *Anatomy of the Law* [Chicago: Britannica Perspectives, 1968]). Dworkin is arguing that change is possible—and inevitable—even in such an artifact.

10. Take as a simple example getting in line for a bus. A just set of rules is created as those who need a ride wait their turn and do not take advantage of the opening created by everyone's being in line to jump on board. Justice comes from our natural need for some cooperative scheme on the basis of which we can act, having some idea of how others shall act. Justice is a set of rules, in short, that regulate the ways in which we relate to one another, ways that may seem contrary to our immediate self-interest but in fact further that self-interest by guaranteeing that others will have reasons not to act for their own gain in social situations provided they have reason to believe you will not act for your own gain in such situations. For an analysis of Hume's theory of justice as a natural social artifact, see my "Hume and the Constitution," in Alan S. Rosenbaum, ed., *Constitutionalism: The Philosophical Dimension* (Westport, Conn.: Greenwood Press, 1988), pp. 31–53, with the caution that this article contains some errors introduced during the printing process.

11. We can cause harm independently of any intent to cause harm. I may grab someone about to fall and inadvertently wrench the person's arm, causing harm, even though I acted with the most benevolent of intents and acted with due care and concern not to cause harm. A harm is simply a setback to one's interests, and such setbacks can occur without intentional action by anyone. A storm may topple a beloved tree in one's yard and so set back one's interest in shade and the beauty of one's yard. For an analysis of harms as setbacks to interests, see Joel Feinberg, *The Moral Limits of the Criminal Law, Vol. I: Harms to Others* (New York: Oxford University Press, 1984), pp. 33ff. I examine the nature of harms in more detail in Chapter 6.

12. Problems can arise that are difficult to solve because their mode of entry into the system is unclear. A recent outbreak of the bacteria *Escherichia coli* in New York City's water supply system, affecting "two distinct areas of Manhattan," puzzled experts. The water entering the system appeared clear of the bacteria, and either "the bacteria entered farther downstream [from the reservoirs] in the system that serves the affected neighborhoods, or . . . trace amounts that are always

present were somehow able to blossom there" (Matthew L. Wald, "Bacterial Taint in Water Supply Baffles Experts," *New York Times* [July 29, 1993], p. A1). One factor complicating the problem is that, as Albert F. Appleton, the city's commissioner of Environmental Protection put it, "Nothing alive should have been there." The bacteria survived "three times the dose that New York City uses to assure safety" (ibid.). As one expert said, "This is not a simple problem with a simple solution" (ibid., p. B2)—a remark on just how complex a mechanism any water supply system is and on just how difficult it can be to ensure a safe water supply. The likely cause, it turned out, is that "the bacteria . . . came from sea gull droppings in a reservoir in Yonkers," always present there, but stirred up by a greater amount of water being flushed through the reservoir because of construction (Matthew L. Wald, "Sea Gulls in Yonkers Are Suspects in Water Contamination Mystery," *New York Times* [August 4, 1993], p. A1).

13. One of the latest problems to plague New York City's water supply is the increase in the number of swans who have "found a home in the Muscoot Reservoir in Westchester County, part of the [city's] water supply system" (Harold Faber, "The Swan, Multiplying Prolifically, Becomes an Ugly Duckling," *New York Times* [February 3, 1993], p. B5). Birds can weigh up to 25 pounds and a flock of them can produce large amounts of feculent matter that, as it settles into a reservoir, can pollute a water supply. The flock in the Muscoot Reservoir number at least 100.

But such problems are minor compared with the major difficulties created by pollution of the watershed for the various lakes and reservoirs that serve a city's water supply. New York City, blessed with an abundance of pure and good-tasting water, faces the equally expensive alternatives of building filtration plants or of controlling development in Catskill and Delaware counties, which would require purchasing large tracts of land.

The City has already lost the pure water that used to come through its Croton system and "has agreed to spend $600 million to filter its water." But although filtered water is safe, it is not necessarily good to drink, and the City can only preserve its good water by preventing development in Delaware and Catskill counties. Former Mayor Dinkins said "it is absolutely asinine to let people pollute the water we're going to drink, and we're not going to have it" (Matthew L. Weld, "A New 10-Year Plan to Prevent Water Pollution," *New York Times* [September 12, 1993], p. A56). The political difficulties of New York City's controlling the development in the two counties of Delaware and Catskill are obvious, and the costs are high no matter what it does (Michael Specter, "New York City Feels Pressure to Protect Precious Watershed," *New York Times* [December 20, 1992], pp. A1 and A46). For some sense of how those the City of New York wants to regulate feel, see George Judson, "Our Town: A Road to Pure Water Paved With Anxieties," *New York Times* (September 28, 1993), p. B8.

14. In the face of a requirement by the Environmental Protection Agency that it filtrate its water, at a cost of "up to $8 billion to build and $300 million to operate," New York City is buying time by buying property in its watershed to keep its water from being polluted initially and by engaging in what can only be described as "manure management" (Matthew L. Wald, "Advocating Vigorous Action on Clean

Water," *New York Times* [June 3, 1993], p. B8). Because the parasite *Cryptosporidium* "is believed to be present in cow manure and can find its way into reservoirs in the runoff from farmyards," and because "a second microorganism, giardia, may also be a [similar] threat," New York City has embarked on an extensive program to decrease the risks of such runoff by encouraging farmers to plow under what manure they use for fertilizer "before rain can wash it away," "plowing across hillsides, instead of up and down, to reduce erosion of soil into reservoirs," and so on (Matthew L. Wald, "Farm Help Keep City's Water Pure," *New York Times* [May 27, 1993], pp. B1 and B4).

15. Other possible problems, considered in the Milwaukee case, were "a cross connection in sewage" and breakage in a pipe somewhere. There are clearly other sources of contamination (Michael deCourcy Hinds, "Survey Finds Flaws in States' Water Inspections," *New York Times* [April 15, 1993], p. A14). See also Sara Terry, "Drinking Water Comes to a Boil," *New York Times Magazine* (September 26, 1993), pp. 42ff. In a letter to the editor, Paul F. Levy, executive director of the Massachusetts Resources Authority from 1987 to 1992, notes that making water "completely safe," as Terry suggests, ignores some major issues that need to be addressed. For instance, it needs to be decided whether public water supplies are to be made completely safe "for immune-deficient people or for the public in general" (*New York Times Magazine* [November 17, 1993], pp. 6 and 8).

16. "Rarely if ever do public water systems periodically survey for the parasite because of the impracticability of the testing procedure" (Lawrence K. Altman, "Outbreak of Disease in Milwaukee Undercuts Confidence in Water," *New York Times* [April 20, 1993], p. C3).

17. It has recently been found that "states fail to check half of the 59,000 large water systems and 20 percent of the 139,000 small systems to make sure that contaminated water, including sewage or runoff with pesticides, does not enter the drinking water," and although states are required to "inspect water systems every three years," some states, such as Indiana, "have inspections at a rate that would require 10 to 16 years to inspect all systems" ("Survey Finds Flaws in States' Water Inspections").

Even where there are inspections, there is still cause for concern because "less than half the inspectors in the country had received any formal training in determining the safety of water, and . . . new inspectors [are] trained by inspectors who had never been trained themselves" ("Outbreak of Disease in Milwaukee Undercuts Confidence in Water"). This failure to train inspectors properly is a particular problem given that such parasites as "*Cryptosporidia* are resistant to chlorination" so that "filtration is the only way to keep drinking water parasite-free." Inspectors clearly missed the problems with Milwaukee's filtration system.

It makes one less sanguine about the problems of ensuring water purity to have those in charge, such as Albert F. Appleton, the commissioner of Environmental Protection in New York City, argue that "the testing system . . . is 'rigged against us'" because it is so stringent as to show even the smallest difficulty (Matthew L. Wald, "Don't Worry but Be Wary: A Fine Line in Water Scare," *New York Times* [August 1, 1993], p. A39).

Of course, the difficulties of ensuring that inspections for public health are themselves done properly are not limited to the water supply. New York City's problems with asbestos in its public schools is a wonderful, though sad, example of the problems that can arise when there is no procedure for inspecting the inspections. It is thought that "the asbestos tests for more than one-third of the city's schools were complete fabrications" (Peter Marks, "Asbestos Tests Were Faked, Officials Say," *New York Times* [August 8, 1993], p. A37), and the school authorities nominally in charge simply paid no attention to possible problems (Robert D. McFadden, "Report Says Officials Ignored the Problem," *New York Times* [August 7, 1993], p. 24).

18. "Lead in drinking water exceeds federally permissible levels in nearly 20 percent of the nation's largest cities," and those levels are set at 15 parts per billion, significantly higher than the World Health Organization's proposal of 10 parts per billion (Michael Specter, "Lead Levels Excessive in Water, E.P.A. Says," *New York Times* [October 21, 1992], p. B8).

19. "Lead can damage the nervous system, the kidneys, and bone marrow. Fetuses and young children are particularly sensitive, and exposure to lead once considered safe is now known to reduce intelligence" (Marian Burros, "Eating Well," *New York Times* [January 27, 1993], p. C4). The evidence seems to suggest that even below the current level at which children are considered poisoned, 10 micrograms of lead per deciliter, they suffer "lasting losses in intelligence" (Jane E. Brody, "Lead is Public Enemy No. 1 for American Children," *New York Times* [November 18, 1992], p. B8). The research that led to this finding has been the subject of some dispute. "In setting its policy on how much lead should be permitted, the Environmental Protection Agency was drawn into the dispute in 1983" regarding the methodology used by Dr. Herbert Needleman, who did the first research. The EPA eventually agreed with Dr. Needleman (Philip J. Hilts, "Hearing Is Held on Lead-Poison Data," *New York Times* [April 15, 1992], p. D28). In any event, whatever the details of the dispute, so long as the research is not shown to be thoroughly mistaken, the form of decision-procedure I will suggest would support taking seriously Dr. Needleman's findings.

20. But not totally. The EPA has released a report saying "that 816 water systems would have to take action to cut lead levels" introduced by the system itself (Matthew L. Wald, "Clarifying a Report on Lead in Water," *New York Times* [May 20, 1993], p. C2). Some water system authorities are adding chemicals to "coat the pipes," and others "are using chemicals to make the water less acidic, so that it picks up less lead" (Matthew L. Wald, "High Levels of Lead Found in Water Serving 30 Million," *New York Times* [May 12, 1993], p. A12). It should be noted that the report covers only "systems with at least 3,300 customers, which together cover about 180 million people" (ibid.).

21. "Is There Lead in Your Water?," *Consumer Reports* (February 1993), p. 75.

22. Ibid., p. 74. Tests recently conducted "by the Environmental Quality Institute of the University of North Carolina in Ashville, N.C., . . . found that faucets manufactured by each of the companies named in a [California] suit leached from 3 to 125 micrograms of lead from a liter of water taken in a single draw" ("Cali-

fornia Lawsuit Says Faucets Leach Dangerous Levels of Lead," *New York Times* [December 16, 1992], p. A22). California law allows an intake of no more than 0.5 micrograms a day. It is some measure of the difficulty of understanding the problems of such matters as lead poisoning that different units of measurement are used by the various official bodies concerned to control it. The main point, of course, is that more lead is in much more of our water than a concern for our health—and our intelligence—allows, and we can tell that by comparing the ratios of what is permissible with what is found.

23. "Is There Lead in Your Water?," pp. 77–78.

24. Burros, "Eating Well," p. C4.

25. What is at issue here are competing explanations for how we can have a system with a design. It may be produced, as I suggest many are produced, with no particular intentions at all in mind regarding the final form such an artifact ought to take. We thus get a system with a design but no designer. Or it may be produced by designers with some intent in mind. If that intent is to cause harm, the latter cast of mind requires a conspiracy theory. We briefly examine this idea of competing forms of explanation later in this section.

26. Anne Raver, "Farms Worried as a Chemical Friend Turns Foe," *New York Times* (February 24, 1992), pp. A1 and 14. The fungicide is produced by DuPont, and the company originally "acknowledged that its product seemed to be at fault and began paying claims—about $510 million." But in 1992, it decided that the fungicide was not to blame, and in response to its action, it was sued (Peter Katel, "The Legacy of Dead Tomatoes," *Newsweek* [August 9, 1993], p. 48). The first case (of over 400) was settled in the second day of the jury's deliberations (Peter Applebome, "DuPont Settles Growers' Fungicide Suits," *New York Times* [August 13, 1993], p. A8).

"In 1989, batches of Benlate powder became contaminated by a weed killer during manufacturing and was temporarily removed from the market. Two years later, the problem was discovered again, and DuPont withdrew Benlate powder from the market" (Keith Schneider, "DuPont Plans Inquiry on Pesticide Crop Damage," *New York Times* [June 17, 1993], p. A23). One problem DuPont has is that some internal documents seem to indicate that "DuPont knew Benlate was contaminated and therefore a plant killer when it recalled the product in March 1991" ("DuPont's Enemy in Lawsuit: Its Own Papers," *New York Times* [August 1, 1993], p. A36). "Taken at face value, the papers indicate that before DuPont ended the compensation program last November, its scientists had nailed down Benlate as the cause of the damage" (Schneider, "DuPont Plans Inquiry on Pesticide Crop Damage.")

27. Other examples can readily be added to the store. For instance, it may seem obvious that providing nesting boxes is one way to increase the numbers of wood ducks in a declining population whose habitat and, so, whose nesting trees are being destroyed. Nothing might seem more innocent. But wood ducks are opportunistic layers. They will lay their eggs in the nest of another wood duck if they can. They usually do not find many such nests because the nests are in old trees, deep in the woods, but new nesting boxes are easily spotted and putting them up has produced quite unintended effects. Some boxes have been found with as

many as fifty eggs in them. The eggs break from the weight, and each is a double loss both because it will not hatch and because the duck, having successfully laid its egg or eggs, will be less inclined to begin its own nest. The result of the introduction of nesting boxes was thus not an increase in the number of wood ducks but a severe decline—exactly the opposite of the intended effect ("Birdhouses for Ducks May Harm Breeding," *New York Times* [May 19, 1992], p. C4).

A more disconcerting example concerns Kemp's Ridley turtle. For years biologists have been digging up their eggs and incubating them and then releasing them to cut down on the loss to predators. But the population has steadily declined despite these efforts. About ten years ago, however, it was discovered that the sex of these turtles is determined by the temperature at which they are incubated and that the researchers have unintentionally been releasing turtles of all one sex. Their mistake would be humorous if it were not so sad. "In the 1940's . . . at least 40,000 nesting turtles could be found on the beach [on which they nest] on one day. Today there are probably only 600 females left and conservationists try to retrieve all their eggs for raising in hatcheries" (Cory Dean, "Their Beaches Eroding, Threatened Sea Turtles Have Few Places to Nest," *New York Times* [March 17, 1992], p. C4).

The turtles released were primarily male, and so two scientists, David P. Crews and Thane R. Wibbels, have patented a process for "increasing the ratio of females to males [so that] the population . . . could be made to increase exponentially rather than geometrically." The process consists of "spotting" "a small amount of the female hormone estrogen in an alcohol solution . . . on the outside of an egg." That "will cause the . . . embryo to develop as a female." The method is claimed to be "much cheaper and more practical" than using incubators—and, hopefully, more effective in increasing the population of Ridley turtles than the old method (Teresa Riordan, "Patents," *New York Times* [April 19, 1993], p. D2). For a discussion of what sorts of pressures sea turtles are under that affect their chances of survival, see "Sea Turtles: In a Race for Survival," *National Geographic* (February 1994), pp. 94–121.

28. William K. Stevens, "River Life Through U.S. Broadly Degraded," *New York Times* (January 26, 1993), p. C1.

29. There are, of course, many examples of acts regarding the environment producing exactly the opposite of an intended end. Consider the introduction of "the opossum shrimp *Mysis relicta* . . . into the great Flathead Lake and associated river systems of Montana." These were introduced as food for the kokanee salmon, which attracted grizzlies and "spectacular concentrations" of bald eagles that gathered in the area on their southern migrations. But "the opossum shrimp consumed zooplankton that formerly were the principal prey of the kokanee. The shrimp were able to avoid becoming the kokanees' new prey by a behavioral adaptation; they migrate to deep waters during the day, thus minimizing their vulnerability to predators" (J. David Allan & Alexander S. Flecker, "Biodiversity Conversation in Running Waters," *BioScience*, Vol. 43, No. 2 [January 1993], p. 40). So the salmon population crashed, and the bald eagles, grizzlies, and tourists went elsewhere.

30. David Hume, *Dialogues Concerning Natural Religion*, Norman Kemp Smith, ed. (Indianapolis: Bobbs-Merrill, 1947), p. 196.

31. The test is a dirty test, however, in part because there is no way to guarantee that one person's observations of what an "objective observer" would note will match up with any other person's observations. We have, in short, no way to guarantee objectivity with this Humean experiment.

But the test is helpful, even if dirty, because it forces us to back off from an ongoing practice and ask about its point or points: Why do we have such a practice? And it forces us to evaluate a practice in terms of whatever its point or points may be. It initiates, that is, a discussion about practices, one that must somehow get started if any practice is to be assessed.

32. Stevens, "River Life Through U.S. Broadly Degraded," p. C1. Patterns of land and water use are more destructive than spills because, it is claimed, they forever alter the habitat. Streams can eventually be restocked if, to concern ourselves with only one life form, the fish are killed in a chemical spill, but fish cannot live when the stream itself is so altered as to be inhospitable.

A recent spill in California is some evidence for this. A pesticide was spilled in the Sacramento River and "killed every living thing in the river." Banky Curtis, of the Fish and Game Department, claimed that "from a fishery standpoint, this river has been eliminated." But, he added, in confirmation of Dr. Allan's claim, "It will probably take quite a bit of time, at least several years, maybe more, to reestablish the food chain." The point is that although over 100,000 fish were killed, "the system is capable of healing itself naturally" because the essential features of the river have not been changed ("Expert Says Pesticide Spill Killed River Life," *New York Times* [July 22, 1991], p. A7).

33. There are six agents of the degradation of the environment, according to Allan and Flecker: "habitat loss and degradation, the spread of exotic species, overexploitation, secondary extinctions, chemical and organic pollution, and climate changes" ("Biodiversity Conservation in Running Waters," p. 35).

Allan and Flecker give a variety of examples of how these six factors affect aquatic habitats, including many in which the harmful effects are wholly unintended. "For example, a variety of native species, including the Sacramento sucker, rainbow trout, California Roach, and threespine stickleback, were displaced from deeper water habitats in the presence of the Sacramento squawfish, a predatory cyprinid introduced into the Eel River of California" (ibid., p. 39).

34. Or to assess any of our practices. Consider, for instance, our health care system. It is complex and complicated, but Hume's quick and dirty test gives us access to it and gives us the capacity to make intelligent and reasonable assessments of its successes and failures.

A stranger dropped into this world might think the system should serve our minimal health care needs—that if one is born, one presumptively wants to live; that if one continues to live, one presumptively wants to live with the least amount of harm to one's body and, if harm occurs, with the least damaging harm; that if harm does occur, one wants the harm relieved, if possible, and at least minimized; that one wants to live into an old age where one has care; that we all want all of these things; and that we want all this without harming other interests that we must consider, for example, trading our financial well-being for our health. But as a stranger dropped into our world would tell us, our current health care system does not satisfy these minimal needs. If it is meant to provide health care to everyone, why are so many excluded? If it is meant to ensure that those who are born survive, why is the infant mortality rate so much higher here than the rates of other industrialized countries (Elizabeth Rosenthal, "In Canada, a Government System That Provides Health Care to All," *New York Times* [April 30, 1991], p. A1). If it is meant to ensure that people will not get ill or that if they do, they will not be much harmed, why are so many resources of the system directed towards relief rather than prevention? If the system were designed to ensure that the elderly are well cared for, why is there little long-term care for the elderly? If the system were designed to ensure health care without harming our other interests, why is ours "the only nation that leaves families vulnerable to medically induced financial disaster" (Erick Eckholm, "Rescuing Health Care," *New York Times* [May 2, 1991], p. B12)?

The health care system undoubtedly has a design, but its design is the result of innumerable decisions made for reasons having, it seems, no real connection to the minimal basic medical health care needs. Although we may explain much of the system by appeal to economic considerations, for instance, just as many try to explain much of law by appeal to economic theory, we cannot explain it all. Only someone mad or with no understanding of how to arrange means to ends would have designed a system that bears so little relation to what prima facie ought to be its ends, and so it would seem mad to suppose that some single vision animates it or that intentions transparently permeate it. It seems unreasonable to claim that its design reflects a designer or a set of designers—patients, health care practitioners, government officials—responding rationally to the most minimally basic needs of health care.

Hume's test thus tells us how the actual design of a natural social artifact may fail to meet its presumed ends and gives us a handle on how to begin to comprehend how it got to be what it is and what one ought to aim at in trying now, self-consciously, to change it.

35. "Less than a year after it started operation, Detroit's giant trash incinerator, the largest in the nation, was abruptly shut down . . . by state regulators. Environmental officials said the behemoth plant, which converts waste into energy, was emitting much higher levels of mercury than are permissible under state law." The problem is that because mercury "vaporizes at a relatively low temperature, [it] eludes capture by some more primitive pollution devices" (William E. Schmidt, "Trying to Solve the Side Effects Of Converting Trash to Energy," *New York Times* [May 27, 1990], p. E5).

36. In Miami, from July 1990 to June 1991, almost one-half of the glass items that were carefully separated by home owners and occupants of apartments from the rest of their trash so that it could be recycled was in fact put in landfills because they were broken in transit ("Glass, to Be Recycled, Showers Florida Dump," *New York Times* [September 12, 1992], p. C9). As we will note, the same sort of problem

has arisen regarding the recycling of newspapers because there are not enough plants capable of turning them into useable paper.

37. Bernard Bailyn, *The Ideological Origins of the American Revolution* (Cambridge, Mass.: The Belknap Press, 1982), p. 95.

38. See on this point Robert Nozick, *Anarchy, State, and Utopia* (New York: Basic Books, 1974), pp. 19ff.

39. One feature of bats that some might think mark the presence of a wonderfully gifted designer is the way in which they are able to send what to them are incredibly loud sounds while not causing damage to their large ear drums, which are designed to receive the sometimes very faint echoes. Bats hear the way we do: "sound is transmitted from the eardrum to the microphonic, sound-sensitive cells by means of a bridge of three tiny bones known . . . as the hammer, the anvil, and the stirrup" (Richard Dawkins, *The Blind Watchmaker* [New York: Norton, 1986], p. 27). So we might suppose that such sensitive ears would be badly harmed by the loud sounds bats emit. But the muscles that are attached to these bones of the ear "contract immediately before the bat emits each outgoing pulse, thereby switching the ears off so that they are not damaged by the loud pause" (pp. 27–28). As Dawkins goes on to remark, the system is like that designed to allow machine guns to shoot "through" the propeller of a plane, "the timing being carefully synchronized with the rotation of the propeller so that the bullets always passed between the blades and never shot them off" (p. 28). As Dawkins goes on in detail to lay out, the bat is a wonderful example of complex design. It is such examples as this that provide the basis for the argument that such a mechanism had to have a designer, and it is the power of evolutionary theory that it can explain how such an incredibly complex phenomenon can come into existence without presupposing a gifted designer.

40. Consider how we treat our lawns. The standard is that lawns be covered with grass, green and well-maintained. Indeed, "in hundreds of communities the failure to mow is punishable by fines" (Michael Pollan, "Abolish the White House Lawn," *New York Times* [May 5, 1991], p. E17). It would never occur to most of us to consider alternative ways of covering the land around our homes—allowing it to turn to meadow, for instance, or planting it wholly in wild flowers—or even question the need for such an expense that our manicured laws require, maintained as they are only through great quantities of water, fertilizer, and weed killers and great amounts of labor ("30 hours [of mowing] for every man, woman, and child") (ibid.). Yet, once questioned, the practice presents some serious problems. Some are obvious—the expense of fertilizers and weed killers, the costs of producing them both to the environment and to us, the loss of "animal habitat," and so on (Malcolm Jones, Jr., "The New Turf Wars," *Newsweek* [June 21, 1993], pp. 62–63). Some are not so obvious. For instance, the dogs of those home owners who use the herbicide 2,4-D are twice as likely to develop lympatic cancer ("Lawn Herbicide Called Cancer Risk to Dogs," *New York Times* [September 4, 1991], p. B7). If one adds up the harms and benefits, it is not at all obvious that green, well-maintained lawns ought to win out as *the* standard or would win out by Hume's quick and dirty test.

For an overview of the environmental issues raised by our having lawns, see F. Herbert Bormann, Diana Balmori, and Gordon T. Geballe, *Redesigning the American Lawn: A Search for Environmental Harmony* (New Haven: Yale University Press, 1993), pp. 86–117.

41. I have in mind here practices such as our health care system. I do not mean to assume that all practices need be assessed.

42. See in this regard Ferdinand de Saussure, *Course in General Linguistics*, trans. and ann. Roy Harris (London: Duckworth, 1983). His general explanatory model for language is Darwinian, and he uses evolutionary concepts to explain such diverse matters as how languages are marked as different from one another (the impossibility of someone from one understanding someone from the other, like the impossibility of impregnation being used as the standard for distinguishing natural species) and how sublanguages are created (through geographical isolation, for instance, just as subspecies of various animals can be formed).

43. Consider the evolution of the rules of courtesy that make it not clearly courteous in some cases, and clearly discourteous in others, for a man to hold open a door for a woman. There has been a sea change in such small courtesies over the past twenty-five years or so, leaving some well-meaning gentlemen stranded. This is the social practice Dworkin examines (see *Law's Empire*, pp. 47ff.).

44. For a brief discussion of some features of this development regarding privacy, see my "The Constitution and the Nature of Law," *Law and Philosophy*, Vol. 12, No. 1 (January 1993), pp. 5–32.

45. An average of 225 tons a day of sludge from New York City is arriving in Sierra Blanca, Texas. The city is the county seat of Hudspeth County, almost the size of Connecticut, with a population of 2,600. The sludge is being dumped there and is considered waste (Roberto Suro, "Texas Town and Fertilizer from That City," *New York Times* [January 25, 1993], p. B2).

But in Lamar, Colorado, farmers are happy to see it. "Your waste comes out here and fertilizes our wheat fields. That helps make some of the bread that finds its way back to your tables," says Douglas Tallman, a farmer in Lamar, and New York City's environmental commissioner, Albert F. Appleton, says, "We've got a hell of a commodity here" (Michael Specter, "Ultimate Alchemy: Sludge to Gold," *New York Times* [January 25, 1993], pp. B1 and B2). Of course, the story of how waste becomes a resource is not so simple or unproblematic as the environmental commissioner and Mr. Tallman's comments may lead one to believe. It is not obvious that the wastes are entirely free of toxic compounds that may waste the soil being fertilized, and it is a decision to be weighed whether the risk of harm is worth the gain. We consider how such decisions ought to be made in later chapters.

46. It should not go unnoticed that the concept of evolution is itself not stable. The traditional model of it as a slow process requiring enormous lengths of time is under pressure from those who think there have been evolutionary bumps that have caused mass extinctions and created such different climactic conditions as to cause vast changes relatively quickly on the cosmic scale. We do not need to take a position on these current debates within evolutionary theory to take as an

hypothesis that natural social artifacts evolve. Even those who think evolution is subject to cosmic bumps believe that the evolutionary process between such cataclysmic events is slow and steady.

47. One could argue that the whole of linguistic philosophy is based on this insight, but it is difficult even to begin to do justice to this claim—or the insight. One should begin with Stanley Cavell's "Must We Mean What We Say? (*Must We Mean What We Say* [New York: Scribner's, 1969], pp. 1–43). But the deeper implications of my claim about this insight are to be found in Wittgenstein (*Philosophical Investigations* [Oxford: Blackwell, 1958]). A thorough analysis of his understanding of how language is learned, for instance, shows that a practice, and *thus* a particular understanding of rules, is built into a use of language. So to appeal to the way we must speak to be understood is to appeal to the norms of what I call a natural social artifact.

48. So overhearing someone say in apparently casual conversation, "Said I to myself . . ." is striking. It is English, and understandable, but not the norm. The inference of the person who heard the remark was that the speaker was an academic, but as Hugo Bedeau has pointed out, a more likely inference is that the speaker knows Gilbert and Sullivan.

49. The following discussion draws from H. L. A. Hart's discussion of the distinction between what he calls internal statements and external statements, the former drawn from within what I would call a natural social artifact, the latter drawn from outside. Hart makes the distinction to make the point about legal systems that their features are contingent and yet are normative for those within them (*The Concept of Law* [Oxford: The Clarendon Press, 1961], pp. 99ff.). The example I later use of stop signs is drawn from his discussion.

As I have noted, a legal system is an instance of a natural social artifact, but it differs from many, if not most, by having a large number of its norms, though clearly not all of them, self-consciously adopted. I say clearly not all because some of the most important norms are never articulated within the system. In Chapter 4, I consider an example of a fundamental norm of our legal system that is not articulated as a norm.

50. The phrase "term of criticism" is Stanley Cavell's, and he uses it to note that various philosophical traditions have their own terms of criticism, tied to what they take to be the most important features of their views and thus to what they take to be the most important failures. One way of putting this that is not too misleading is that the terms of criticism mark the ways in which something fails to match up to what a particular view takes to be a paradigmatic example of something. So we may investigate the normative force and features of a natural social artifact by exploring its terms of criticism or its paradigmatic instances. See Stanley Cavell, *The Claim of Reason* (Oxford: Oxford University Press, 1979), pp. 118ff.

51. John Stuart Mill discusses how it is that we enforce the internalization of norms—though not in those terms—when he talks about when it is appropriate for someone to be "punished by opinion, though not by law" (*On Liberty* [Indianapolis: Bobbs-Merrill Company, Inc., 1956], p. 92). He says that "though doing

no wrong to anyone, a person may so act as to compel us to judge him, and feel to him, as a fool or as a being of an inferior order" (ibid., p. 94). Someone "who shows rashness, obstinacy, self-conceit—who cannot live within moderate means; who cannot restrain himself from hurtful indulgence; who pursues animal pleasures at the expense of those of feeling and intellect—must expect to be lowered in the opinion of others" and to be criticized accordingly (ibid., p. 95).

52. The claim that norms *primarily* depend for their normative power on our internalizing them is not one I cannot properly support here and is, in addition, pregnant with riches for moral analysis. Let me refer to just one here. It may, for instance, be more efficient that children be trained in such a way that they regulate their own behavior without external monitoring, but it is not obvious that it is morally preferable. Is it morally better that someone know about the options and choose, or be so reared that such options never occur to them—at least as serious options? Answering this question would lead us into some deep differences between Kant's theory, for instance, and Mill's.

53. A company's worth, as indicated by the accountants charged with doing its books, is an artifact of whatever accounting rules are used. These are themselves not determinative of the worth of the company in part because they leave much leeway for interpretation but also because the worth they determine may bear no relation at all to the actual worth of the company—what would be left in hand if its assets were sold and its debts paid.

A dramatic example is the British company of Polly Peck International, a food and consumer electronics company. In September 1990 it showed profits of 110.5 million pounds, with various interests in other companies valued at 660 million pounds, and with shareholders' funds of 932.7 million pounds. By the end of October, it had gone under, and "the results of an insolvency study by [the accounting firm of] Coopers & Lybrand were revealed allegedly showing that an immediate liquidation of Polly Peck would produce a deficit for shareholders" of 384 million pounds (Terry Smith, *Accounting for Growth* [London: Century Business, 1992], pp. 7–9). Smith details other instances as well, including Robert Maxwell's companies.

54. This line of reasoning about one of the major causes of the savings and loan crisis is Lawrence J. White's in *The S & L Debacle: Public Policy Lessons for Bank and Thrift Regulation* (New York: Oxford University Press, 1991). As he puts it, "the bank and thrift regulatory (and deposit insurance) regimes of the United States are based on a fundamentally flawed information system. That information system—the standard accounting framework used by bank [sic] and thrifts—looks backward at historical costs rather than at current market values. It does not yield information that would allow regulators to protect the deposit insurance funds properly. . . . The revamping of this accounting framework—a switch to market value accounting—is the single most important policy reform that must be accomplished" (pp. 3–4). This accounting framework "gives depository managers a tremendously valuable 'option': They can sell their 'winners' to show gains, while hiding their 'losers' by continuing to carry these latter assets in their portfolio at original costs" (p. 226).

55. I do not mean to pick on accounting here. The problem is a general one in

professions, for the mode of reasoning into which a professional is trained may well include a form of decision making that can blind one to other forms.

The *Challenger* disaster is marked in part by the inability of the person who effectively ordered the flight to understand why he was blamed for its failure. He was an engineer, and as an engineer, his judgment was that they should not fly. Although no evidence had been gathered of what would happen to the O-rings when the temperature fell below 40 degrees Fahrenheit, what evidence they had gathered suggested that the rings would become more and more brittle. An engineer is trained to be risk-averse, and the evidence thus would suggest to such a person, with such a stance regarding risk, not to fly. The man who ordered the flight was an engineer, but also a manager, and when he reported that in the judgment of the engineers *Challenger* should not fly, he was asked to put on his "manager's hat." Managers are trained to make decisions by balancing costs and benefits, and given, among other things, that NASA's budget was up for a vote in Congress the next day and that NASA had done poorly with its space shuttle program that year, the benefits, on this view, outweighed the costs. So the engineer, with his manager's hat on, decided that *Challenger* should fly. The mistake he made, one he did not realize he was making apparently, was the mistake of not backing off and deciding which was the appropriate form of decision making in this case, the risk-averse stance of engineers or the cost-benefit analysis of managers. For a careful and good analysis of this problem, see Michael Davis, "Explaining Wrongdoing," *Journal of Social Philosophy*, Vol. 20 (Spring/Fall 1989), 74–90.

56. "In the age bracket between 45 and 64, before women qualify for Medicare, only 55 percent of working women have health insurance provided by their own employers, as compared with 72 percent of men in the same age group" (Felicity Barringer, "Study Says Older Women Face Insurance Gap," *New York Times* [May 7, 1992], p. A19). Women are more likely to work part-time, in low-paying jobs, and in small businesses that cannot afford health care.

57. The infant mortality rate for blacks is double that for whites, for instance, and blacks have higher death rates for "many of the major causes of death" ("Health Report Says Racial Disparity Lingers," *New York Times* [September 16, 1993], p. A16).

58. Of the 13.1 percent of Americans not covered by health insurance in 1988, 26.5 percent were Hispanics, 20.2 percent were black, and the dividing line for coverage is the 35th birthday. It is then that the percentage not covered falls below 13.1 (Tamar Lewin, "High Medical Costs Hurt Growing Numbers in U.S.," *New York Times* [April 28, 1991], p. A14).

59. Sick newborns get treatment even if the relevant parent does not have health care insurance, but those whose parents who do not have health insurance receive 28 percent less care and are discharged two and a half days earlier, on average, than sick newborns whose parents have health insurance or Medicaid ("Study Ties Hospital Care of Newborns to Coverage," *New York Times* [December 18, 1991], p. A26).

60. The argument of this book presupposes a conception of morality in which

we can cause harm to others by setting back their interests without any morally relevant intent.

Many ethical issues in public policy and in business and the professions in fact arise because of what we might call *design* flaws. A natural social artifact, such as the internal structure of a corporation, for instance, or a law firm, may have, as an unintended by-product, untoward moral effects. A system is poorly designed that, for instance, puts someone in the position of having to choose, to correct a harm, between blowing the whistle publicly on his or her company and not blowing the whistle but countenancing the wrong. Having to face such a choice is a harm because, among other things, the employee will be harmed, regardless of what choice he or she makes.

If that choice comes about because of a natural social artifact that could be different from what it is, there is a moral obligation to determine if such a sort of harm is inevitable, given the nature of the ends the particular social artifact is meant to answer to, or whether some changes in the artifact will remove, or at least mitigate, those harms without causing new ones at least as severe. It is a moral wrong knowingly to support an error-provocative design, especially when the errors cause harm to innocents, as in our health care system's providing less care for those sick newborns whose parents do not have health insurance or Medicaid.

61. There is much to be said about (1) what natural social artifacts are produced by which causes, moral and otherwise; (2) what natural social artifacts produce which effects, moral and otherwise; (3) what values and moral implications underlie and are furthered by the particular natural social artifact of waste management we now have; and (4) what values and moral implications could be furthered by a different waste management system.

The first two are general enquiries beyond the scope of this narrative, but I briefly remark on the methodology I think they entail in Chapter 4. I am hawking no general overarching thesis about how natural social artifacts come to be formed, and I am skeptical of there being any one kind of cause that permeates and forms the whole of a natural social artifact. One can, for instance, provide an economic analysis of much of the law, explaining decisions in terms of the economic benefits and losses of the various parties involved, including the system itself, but one risks leaving out much that is in the law as well as distorting things by emphasizing one kind of causal factor to the exclusion of others.

I have little to say about points 3 and 4. I think that a particular way of reasoning underlies our current system, one encouraged by political factors, by economic considerations, and no doubt by many other causal factors. My interest is in that mode of reasoning and in some of the changes that will occur if that mode of reasoning is displaced by the one I think is appropriate (see Chapter 6).

For those interested in pursuing some of the broader issues raised by points 3 and 4, see Mary Douglas and Aaron Wildavsky, *Risk and Culture* (Berkeley: University of California Press, 1983), esp. pp. 102ff.). See also Stephen Cotgrove's *Catastrophe or Cornucopia: The Environment, Politics and the Future* (New York: John Wiley & Sons, 1982).

3. Knowledge and Doubt

1. The fight over a new incinerator for medical waste in the Bronx is a classic instance of this sort of dispute. Those in favor claim that it is "the safest plant money can buy," and those opposed argue that no one can know that, that the plant is advertised as a "pilot program," and that "even if no smoke is visible the plant will release dangerous levels of lead and other heavy metals as well as toxins like dioxin" (Ian Fisher, "Builders and Foes Using Bronx Incinerator as Test," *New York Times* [September 8, 1992], p. B3). The large type in this article says, "Each side believes it alone knows what is best for the public."

The concerns about the safety of the plant do not arise simply from a disagreement about how to weigh risks but from disagreements about how much risk there will be. For instance, it matters who operates the incinerator because, presumably, someone with knowledge about the dangers of medical waste, such as a hospital, might be more cautious in handling the waste, both before and during incineration, than someone not so knowledgeable. One source of concern is who will operate the plant, the hospital or "RemTech [the company that built it] or some outfit under contract to RemTech" (quoting the Bronx Borough President, Fernando Ferrer, in Dennis Hevesi, "Bronx Foes Try to Stop Medical Incinerator," *New York Times* [November 2, 1991], p. 28).

The fight is of some importance because the plant has already been built, and opponents are trying to stop its becoming operational. If they succeed in stopping it at such a late date, few are going to be willing to invest in future plants.

2. The Palisades Nuclear Plant on the shore of Lake Michigan will run out of storage space the next time it refuels, and it has prepared for that by purchasing "8 cylindrical casks, 16½ feet tall and 11 feet in diameter," with a steel "honeycomb" inside "which holds 24 assemblies or 30 tons of spent fuel" (Matthew L. Wald, "Battling Nuclear Waste in Michigan," *New York Times* [December 8, 1992], pp. D1 and D13). These casks are cooled by the natural circulation of air rather than by the "complex mechanical systems [used] to cool and filter the water" that surrounds spent fuel, and so they can just sit out in the open with "radiation levels near the cask . . . low enough for quick inspection" (*ibid.*, p. D13).

The fight against this method of storage is being led by Dr. Mary P. Sinclair, and the dispute is one of those strange ones where each side trades charges, but the charges seem to slip right past, without doing damage. Palisades concedes that there is a risk but thinks it not high enough to cause problems; Dr. Sinclair concedes that something must be done with the spent fuel but thinks the casks are too risky.

3. A person's *perception* of a risk obviously need not be identical to what a proper risk assessment tells us the risk is, but, again, I would suggest that it is a mistaken procedure to presume that one party to such a dispute *must* be perceiving risks that are just not there. The rhetoric of such disputes guarantees that each party will claim of the other that the risk it purports to perceive is nonexistent. But rather than get caught up in what the logic of the disagreement forces

us to do, we should rather presume the rationality of both parties and attempt to uncover what it is about how each conceptualizes the situation that makes their risk assessments reasonable—and incompatible. If we discover that one party has an incoherent conception, or if we can in no way make its assessment reasonable given its conception, it will be time enough to consider accusing that party of a misperception or of essential irrationality.

4. What proceeds is informed by what I take to be a healthy skepticism about the objectivity of risk assessments. See in this regard especially K. S. Shrader-Frechette, "The Conceptual Risks of Risk Assessment," *IEEE Technology and Society Magazine*, Vol. 5, No. 2 (June 1986), reprinted in Albert Flores, ed., *Ethics and Risk Management in Engineering* (Latham, Md.: University Press of America, 1989), pp. 73–90. Flores's volume contains a number of articles that provide a general background to the issue.

Despite appropriate skepticism, I recognize that risk assessment has its place, properly understood and constrained. See in this regard especially Michael E. Kraft, "Analyzing Technological Risks in Federal Regulatory Agencies," in Michael E. Kraft and Norman J. Vig, eds., *Ecology and Politics* (Durham: Duke University Press, 1988), pp. 184–207.

5. I expand on this point at the beginning of Chapter 7.

6. Julian L. Simon and Aaron Wildavsky, "Facts, Not Species, Are Periled," *New York Times* (May 13, 1993), p. A23. The World Wildlife Fund is quoted in this OpEd piece.

7. It has been claimed that the form of the deforestation matters significantly because "it more than doubles the area of ecological damage caused by deforestation itself" (William K. Stevens, "Loss of Species Is Worse than Thought in Brazil's Amazon," *New York Times* [June 29, 1993], p. C4). This projection, of course, is as much a result of presumptions about how to determine loss of species as any of the others, but the reason for citing it and similar projections is that the *data* cited for them are worrisome—whatever the projections one makes from it.

8. The variety of factors that can make a difference in risk assessments is enormous, and some of the difficulties would be funny if they were not so sad. New York City had a trash strike in 1990 because of a 1988 decision to set the dumping fee at Fresh Kills at $40 a cubic yard of trash, up from $18.50. The aims were to encourage recycling and to slow the loss of landfill space. Those who set the new fees succeeded beyond their wildest dreams. The effect of the increase was to drive nearly all the commercial trash out of the city, over 10,000 tons a day, to less expensive landfills in states such as Pennsylvania and Ohio. The city lost $100 million in dumping fees a year, and the trash strike came about because local haulers wanted to be able to charge more because they, now going out of state and facing increased dumping fees from states unwilling to take New York trash, faced increased costs.

Another "consequence of the increased dumping fee . . . has been the growth of transfer stations. Before the garbage is trucked out of state, private haulers often bring it to these local sites so recyclable material can be removed and the rest compacted for shipment. After the dumping fee at Fresh Kills rose, dozens of these stations opened almost overnight in the Williamsburg and Greenpoint sections of Brooklyn" (Alan R. Gold, "Ripples from a 1988 Rule Led to Messy Trash Strike," *New York Times* [December 12, 1990], pp. B1 and B3). The residents of these areas complained of the odors, the truck noise at all hours, and the vermin attracted by the garbage.

This is a rather nice example of an unintended consequence, for it illustrates how trying to solve one problem can cause another that produces, as this one did, significant political nightmares for those who made the original decision. It is unlikely ever to have occurred to those who recommended increased fees at Fresh Kills to help prolong its life and provide space for the garbage of their constituents that they would pay the political price caused by having that garbage dumped at local sites within their districts while awaiting transport to out-of-state sites.

Such implications may seem obvious after the fact but are hard to predict with the kind of certainty one would need to convince those who must set public policy.

9. The original estimate in 1987 for the Eurotunnel, for instance, was 4.87 million pounds. By April 1990 that estimate had already been raised to 7.5 billion pounds, and it has since gone higher still (Steven Prokesch, "Eurotunnel Cost Estimate Up Again," *New York Times* [April 24, 1990], p. C20). For an excellent discussion of how it is that the costs for public projects are always underestimated and the benefits always overestimated, see Martin Wachs, "Ethics and Advocacy in Forecasting Public Policy," *Business and Professional Ethics Journal*, Vol. IX, Nos. 1–2 (Spring–Summer 1990), pp. 141–58.

10. Richard Cook, *Solid and Hazardous Waste Incineration: An Analysis for Citizens and Policymakers* (March 1987), unpublished, pp. 32–33.

11. Ibid., pp. 21–22.

12. Willim E. Schmidt, "Trying to Solve the Side Effects of Converting Trash to Energy," *New York Times* (May 27, 1990), p. E5.

13. Drift-net fishing is used extensively by the Japanese, among others. The nets are 30 miles long and made of nearly invisible plastic. It is claimed that they kill unnecessarily large numbers of marine life that are not commercially viable or are not supposed to be caught. In response to a call for an end to such fishing, "Japanese officials at the United Nations said they would not cut down on drift-net fishing until there was thorough documentation that the nets were causing marine destruction" (Timothy Egan, "Citing Data on Damage to Pacific, Groups Seek Drift-Net Fishing Ban," *New York Times* [November 14, 1989], pp. A1 and A12).

Even after the United Nations banned the use of such nets, it is still being argued that the ban "was a political overreaction based on flawed data." In short, more information is needed (David E. Pitt, "Fishing Fleets Are Pulling in Their Giant Nets," *New York Times* [January 9, 1994], p. A11).

It should be noted that the ban is not fully effective. The Italians have continued using nets, setting "4,500 miles of nets each night during the fishing season— 'enough to span the entire Mediterranean twice,' said Assumpta Gual of Greenpeace" (ibid.).

14. In the Spring of 1989, "five years after F.D.A. scientists concluded that Red Dye No. 3 was a carcinogen, the agency served notice that it was preparing to outlaw the additive. But then Representative Vic Fazio of California, sensitive to the

interests of fruit growers, tossed a monkey wrench. He inserted language in the agriculture appropriations bill instructing the F.D.A. to put off action pending further study" ("Stop Dawdling on Dye No. 3," *New York Times* [November 6, 1989], A22). Senator Wyche Fowler of Georgia blocked the move.

15. It is not always just *politically* advantageous to ask for more information. The demand for additional information before acting can serve some other end than the political end of delaying a decision.

It is claimed of the British medical system that one way in which costs were kept down was to insist "upon extremely high standards of proof before acknowledging the effectiveness of a medical procedure" (Robert Baker, "The Inevitability of Health Care Rationing: A Case Study of Rationing in the British National Health Service," in Martin A. Strosberg, Joshua M. Wiener, and Robert Baker, eds., *Rationing America's Medical Care: The Oregon Plan and Beyond* [Washington, D.C.: The Brookings Institution, 1992], p. 213). Physicians cannot order a procedure that has not met the standards set, and setting standards extraordinarily high guarantees that patients can be provided with what are perceived to be adequate medical services, that is, nonexperimental procedures and services "proven to be efficient," without increasing costs beyond what a limited budget will allow.

Appealing to such an epistemic condition allows those in charge respectable cover for what may be, in some cases at least, questionable medical practice. So it is not just politicians who may have a stake in playing the epistemological card.

16. For a thorough and careful statement of this procedure, see Michael D. Resnik, *Choices: An Introduction to Decision Theory* (Minneapolis: The University of Minnesota Press, 1987), p. 32.

17. Ibid., p. 27.

18. If we do gamble in checkers, it could be on whether our opponent will note the significance of some move that we must chance because the odds are otherwise against us. We might be in a situation where the chance of losing is so high that we risk a move we otherwise could not justify. In short, gambling has its place in checkers, but it is a place circumscribed by features of the game that make it generally inappropriate. One does not *gamble* that an opponent will not notice one's putting a piece directly in line to be jumped.

19. This presumes, among other things, that the person has not artificially limited the amount of information he or she has and that the information available is not so limited that it would determine almost any decision, but is, in fact, a fair amount.

The question of what form of criticism is appropriate for someone who has limited the amount of information he or she has—to ensure deniability, for instance, in a sensitive political matter such as the Iran–Contra affair—is a difficult matter. Is there a moral obligation to obtain what information we can? And if so, when is there such a moral obligation? When we are ultimately responsible for whatever decision is made? Is there an obligation of reason so that failing to obtain as much information as we can convicts us of being unreasonable? These are not easy questions, and no doubt they require different answers depending on what decisions are at issue.

It is equally difficult to know how to answer the questions of how much information is enough—a "fair amount"—when we know we do not have it all and of how much is enough when we cannot know how much we have because we do not know how much would be all. I have some suggestions to make about the answers to these questions in Chapter 6, but, again, I do not think there is some general rule appropriate here. How much information is enough will depend on the case at hand, as the examples of gambling, checking one's checkbook, and so on illustrate.

20. In an editorial about the failure of the White House to act on carbon dioxide emissions, the *New York Times* said, "When proof of the warming finally arrives, the climate will already be locked into a significant temperature rise, one that could alter crop patterns and sea level. Washington should already be taking out insurance" ("Hot Air and the White House Effect," *New York Times* [November 24, 1989], p. 22).

21. As I have noted, an omission to act can have (moral) significance. One can fail to act in at least two different ways. There is no moral difference between the two ways in a situation where you have all the relevant knowledge one can have, a capacity to act, the desire to help, and a professional obligation to act.

22. We have long been told to forgo butter, with its saturated fats, and eat margarine because it is made from "partially hydrogenated vegetable oils made from soybean and corn oils." But "new data show that these oils—found in margarine, vegetable shortening and a host of products ranging from doughnuts and pies to cookies and crackers—may also cause heart disease" (Marian Burros, "Now What? U.S. Study Says Margarine May Be Harmful," *New York Times* [October 7, 1992], pp. A1 and C4).

To take another example, nothing could seem more healthy than a glass of milk fortified with vitamin D to help prevent rickets in the young and weak bones in the old. But too much vitamin D can be equally dangerous, and recent tests indicate that all too often we get too much ("High Levels of Vitamin D in Milk Are Found Hazardous in a Study," *New York Times* [April 30, 1992], p. A9).

23. Cook, *Solid and Hazardous Waste Incineration*, p. 14.

24. Ibid., p. 15.

25. The buildup of carbon dioxide in the atmosphere has led to a variety of inconsistent predictions about what will happen. The system is complex enough that we ought to wonder about our ability to predict, and as Dr. Wallace S. Broecker is quoted as saying, "My feeling is that we overestimate our ability to predict. Many of the things that are going to happen to the planet will be surprises, like the ozone hole over Antarctica. Therefore, we should be much more careful about what we are doing and much more observant of how the system works" (Philip Shaberoff, "Cloudy Days in Study of Warming World Climate," *International Herald-Tribune* [January 19, 1989], p. 8).

26. See Wade L. Robison, "Philosophy and Public Policy: A Look at PBB," *The Michigan Connection* (Letter of the Michigan Council for the Humanities), Spring 1977.

27. The difficulties with the incineration plant in East Liverpool, Ohio, are an instance of this problem. It has been in the works for over ten years, and over $160

million have been spent by a private company that manages it. There are serious questions to be asked about how it came to be built, with allegations of governmental abuse in approving the project. But the main source of dispute concerns whether its operation will cause great harm and whether there is an alternative. One claim of those who support it is that it will produce less pollution than "the steel and chemical plants and coal-burning plants up and down the river." The implication is that it is just one more, quite minor, source of pollution, and such an argument would need to be buttressed by a claim that not to have an incinerator would mean that more pollution would occur. The argument of those who oppose the plant's operation is that it will cause great harm. Everyone concedes that "the plant will release thousands of pounds a year of pollutants like mercury, heavy metals, lead and sulfur dioxide into the air of the Ohio River Valley." The claim is that the air will be turned "into a toxic waste dump" and that the residents of the valley will be harmed. "The plant is built on a flood plain near a residential neighborhood in a river valley known for its stagnant air" (Keith Schneider, "Gore Says Clinton Will Try to Halt Waste Incinerator," *New York Times* [December 7, 1992], p. D9).

Whatever the detailed merits of this dispute, it should be noted that the form of the debate for those most affected is what we should expect, namely, that we must show great harm to prevent its operation. So it is perfectly predictable that when problems occur with the plant, as when "an equipment malfunction released contaminants," the operator of the plant will issue a statement that "there was no health risk to employees or the public" ("Ohio Incinerator Halted after Malfunction," *New York Times* [December 12, 1993], p. A36). Hovering in the background of this dispute is the issue of what will be done with all the hazardous waste that would have been incinerated in this plant if it does not continue to operate.

28. Beginning January 1, 1992, the citizens of Massachusetts can no longer "send fallen leaves and small branches to dumps and incinerators" ("An End to Dumping Massachusetts Leaves," *New York Times* [January 2, 1992], p. C9). The move is part of the state's plan to cut its trash volume in half over four years by, in this case, preventing leaves and small branches entry into the flow, and it clearly represents the trend.

29. It is easy to find examples of such conflicts and difficult to know how to respond in any sort of general way to the moral problems they present. Hungary has recently issued a decree subjecting all cars to annual pollution inspections and permitting police to pull over obvious polluters for random checks. "The main culprits are the former East bloc's fume-spewing specialties: the Trabant and Wartburgs, with engines produced in what was once East Germany that emit high levels of hydrocarbons." But, as one former official put it, "People will say, look, I didn't want to buy a Trabant either but that was all there was to buy when I could buy a car." It seems obvious that the new practice is better than the old, but, still, the conceptual space is there for the claim that the old practice has some normative force. In this case that claim gets played out in terms of those guilty of polluting under the new practice claiming that they are not fully responsible for what they did and thus for what they are now doing, and that claim has *some* weight in any

calculation. After all, those who purchased Trabants and Wartburgs are correct: they had no choice if they were to have a car, and they now have an investment the new practice taxes.

The issue is complicated by our sense that those who purchased such cars knew they were purchasing cars that pollute, by the realization that the offending cars belong to the working classes, not the rich, who can afford "Mercedeses, BMW's and the like," and by the emission standards that were previously in force containing a specific exclusion for cars manufactured in Eastern Europe. So to ban such cars would be particularly unfair: It would penalize the poorer members of society for purchasing the only cars available to them and for doing what they might have known would pollute, but what was permitted by the government. It is difficult to be judicious in such a complicated moral situation (Celeste Bohlen, "On a Clear Day, You Can Glimpse the Tow Trucks," *New York Times* [January 12, 1991], p. A2).

30. A case in point concerns the decision in December 1991 to restart the K-reactor at the Savannah River Site in South Carolina. The reactor had been closed for almost four years "because of safety concerns that led to the permanent shutdown of the other two reactors at the site." The K-reactor was started again only with the assurance that it was safe. But "during the beginning of the startup procedures, about 150 gallons of highly radioactive cooling water containing tritium leaked into the adjacent Savannah River through a crack the size of a pencil lead. The leak went undetected for two days because the official in charge of authorizing transportation of water samples to a nearby laboratory was out with the flu" (Peter Applebome, "Anger Lingers after Leak at Atomic Site," *New York Times* [January 13, 1992], p. A7).

The official response was that no harm was done. Secretary Watkins flew to the site and declared, "There was never any danger to the public," and Peter M. Hekman, Jr., the "manager of the Savannah River Field Office for the Energy Department, said: 'We are talking about one-hundredth of what a person can have in a year's time from tritium. This is not the first time, nor even the largest dose of tritium released by us'" (ibid., p. A7).

But these official responses completely miss the point. One concern expressed was that the leak was missed. As the local State Representative put it, "Two years ago they had a similar spill, and an internal report said they had to start monitoring it in a different way. Two years later, they have never done it" (ibid., p. A7). Having made a commitment to change things to ensure the safety of those on the river and nearby, the plant managers have a system designed to fail if one person is out with the flu. It is hard to trust one's well-being to such incompetence. Even if this leak caused no harm, one can have no assurance, given such a record, that there will not be more.

The other concern is that the plant was restarted with assurances that it was safe despite the problems with the other reactors on the site. But it was not safe, and as one man put it at a hearing, to loud applause, "I don't believe a thing that you're saying" (ibid., p. A7). That is a rather dramatic way of saying that trust has been lost—although the point might be better put by saying we do not know which assurances to believe and which not to believe. Mr. Watkins can assure the local residents as much as he wants; they have good reason not to trust any such assurances.

This no doubt will frustrate Mr. Watkins, who is presumably acting in good faith. But it is an object lesson in how even the most well-intentioned individual can lose an audience's trust and of the costs of the loss of trust: Nothing he or anyone else can say will now reassure.

31. Such assurances and their subsequent failures seem common in environmental matters. When the North River sewage plant in New York City was built, it was presented as a model of how to build a plant consistent with the needs and desires of the community. Design changes were added that allowed a $130 million state park on top of it, for instance. But the neighborhood was drenched in a stench even before the plant was fully operational despite assurances that the plant was state-of-the-art and would be so clean that no one would know it was there. It is claimed by residents that "on the hottest summer days . . . even air-conditioning 20 blocks away does nothing to dissipate the smells" (Michael Specter, "Stench at Sewage Plant Is Traced and Repair Is Put at $50 Million," *New York Times* [April 17, 1992], p. B2).

It was claimed that the stench at the plant would disappear when it was completed, that the stench "may have been due to open sewers nearby," and then that "the smells were related to the transfer of sludge . . . to barges in the Hudson River." In short, the response was that "there wasn't even a problem there" (ibid., p. B2). But it is going to cost up to $50 million just to reduce the smell so that, only on the worst of days, it is claimed, will the nearby residents be bothered.

The projected $50 million has jumped to $55 in six months since the settlement was reached, and that does not take account of the legal costs—including at least the $1.1 million to settle "a lawsuit filed by two groups and several East Harlem residents" (Richard Pérez-Peña, "City Settles Suit on Odors at Harlem Site," *New York Times* [January 5, 1994], p. B1).

32. It is somewhat ironic that the attempt to introduce "free-market efficiencies to a command-and-control system" of utilities by allowing them to trade emissions allowances among themselves has faltered because the utilities are risk-averse (Matthew L. Wald, "Risk-Shy Utilities Avoid Trading Emission Credits," *New York Times* [January 25, 1993], p. D2). They are concerned in part that buying emission credits may make them look as though they are willing to pollute, but the main problem seems to be that they are more concerned to comply with the law than to profit. The consequence is that "nearly 20 percent of the affected total [of coal-fired power stations] have decided to build 'scrubbers' without a clear idea of how much it would cost to meet their obligations by buying the allowances instead" (ibid., p. D2). The risk-averse attitude affects the engineers and planners who must buy the emissions allowances. As Don Justin Jones, "the managing director of a consulting firm that specializes in utilities and environmental issues," puts it, "If you screw up you get shipped off to count transformers" (ibid., p. D2).

33. This is to say at least that the empirical evidence in regard to environmental issues is decidedly mixed. See on this issue Ian Hacking, *The Emergence of Probability* (Cambridge: Cambridge University Press, 1986), p. 78.

34. Examples may be multiplied almost endlessly regarding other environmental matters. Consider, for instance, the following two examples about gasoline. Gasohol, which is gasoline with 10% grain alcohol, is supposed to cut carbon monoxide but may, in addition, cause more smog. One study indicates that gasohol would cut carbon monoxide by 25% but would produce 8 to 15% more nitrogen oxides, thus producing about 6% more smog (see Matthew L. Wald, " 'Gasohol' May Cut Monoxide but Add to Smog, Study Says," *New York Times* [May 9, 1990], p. A1).

Another example may be drawn from the effects of the new "clean" gas, the reformulated gas that is meant to make new cars run clearer. That gas makes 1983 through 1985 models run significantly dirtier. The new clean gas reduces what are called aromatics and thus reduces the emission of hydrocarbons, which is a main culprit in the production of smog, but although that reduction produces a 6% fall in emissions from new cars, it produces a 9% rise for cars made between 1983 and 1985. It is not clear why (Matthew L. Wald, "For Some Cars, 'Clean' Gas Is Found to Raise Emissions," *New York Times* [December 19, 1990], p. D3). The same study also showed, more generally, that "changing the gasoline recipe to cut emissions for one pollutant increases a car's production of other pollutants" (ibid., p. D1).

35. It is one of the paradoxes of environmental studies that one of the most effective methods for checking ground-water flow is to trace such contaminants as tritium, krypton-85, and chlorofluorocarbons. Because we know when and in what amounts these have been released into the atmosphere—tritium in atmospheric nuclear testing and kryton-85 mainly "during the processing of fuel rods for nuclear power plants"—and we know their half-lives, we can track them accurately in ground water (Tim Hilchey, "Pollutants Traced in Groundwater," *New York Times* [November 30, 1993], p. C5).

36. It is some evidence both that there are no free lunches in responding to environmental problems and that we do not know what we thought we knew that recent studies indicate that in reducing "emissions of airborne sulfates that cause acid rain," we have also inadvertently reduced "airborne alkaline particles that help neutralize the acid. The alkaline reductions," it is claimed, "offset the cuts in sulfates by 28 percent to 100 percent in different places" (William K. Stevens, "New Study Says Acid Rain Efforts Inadvertently Undercut Own Goal," *New York Times* [January 27, 1994], p. C20). The Clean Air Act amendments of 1990 "required industry, by the year 2000, to cut sulfur dioxide emissions by 12 million to 14 million tons a year, or about 45 percent of the emissions in the mid-1980's," but this finding will increase pressure to cut the emissions even more to make up for the offset of the reductions by the attendant reductions in alkaline particles (ibid.). See also Eville Gorham, "Neutralizing Acid Rain," *Nature*, Vol. 367 (27 January 1994), p. 321.

4. Who Decides?

1. For instance, John S. Dryzek claims that the primary focus of concern ought to be the collective choice mechanism in its entirety and that "the nature of the collective choice mechanism will largely determine the kind of world that ensues." He

adds that "it is clear that our collective choice structures have led us into situations which are in nobody's real interest and which very few people wanted, the threat of nuclear holocaust being the most obvious example. Public policy outcomes," he continues, "often bear only a remote connection to the intentions of 'policymakers'" (*Rational Ecology: Environment and Political Economy* [Oxford: Blackwell, 1987], p. 9). His claims are similar to mine about how natural social artifacts produce unintended effects, and one way to put my claims in this chapter vis-à-vis his is that one feature of our "collective choice mechanism" skews our choices, producing results that no stranger to this world would presumably think we would intend.

2. Picking this vantage point from which to view the decision-making process best highlights the particular ways in which that process affects the choices for environmental issues, but I presume that other vantage points—special interest groups, "the citizen," big business—would have other advantages and realize that in choosing such a vantage point, I may be myself skewing the vision and making certain aspects of the decision-procedure appear more significant than in fact they are. The only way one might even begin to be sure of the real weight of such factors would be to lay out the entire natural social artifact that is our political system and see where and how these factors work. I do not think anything I say will be inconsistent, or even problematic, given such a wider view, but testing that hypothesis will have to wait until that much longer story is told.

3. I should add "generally," for it is not true that the standard practice of any profession is immune from difficulties, and we may well criticize a physician, say, for acting as any physician would act if, as it turns out, acting that way causes more harm than some acceptable alternative. Such a problem creates dilemmas for professionals, but that is another issue.

4. The example he gives is persuasive. The difference, he argued, between Poland and Venice was that the princes in Poland all had independence from the crown. They had their own castles and lands, and if their interests conflicted with those of the crown, they suffered no great immediate harm if they went their own way. Poland thus lacked a powerful central government and any subsequent unity and was always open to dismemberment by its neighbors. In Venice the power of the merchant princes depended on the power of Venice as a whole. Every Venice merchant's self-interest was directly tied to the common interests of all of Venice: As its fortunes rose, their fortunes rose, and as its fortunes fell, theirs fell. This fundamental difference in the ways in which the political structures were organized explains much, Hume thinks, about the history of these two very different political entities (David Hume, "That Politics May Be Reduced to a Science," *Essays: Moral, Political, and Literary*, Eugene F. Miller, ed. [Indianapolis: LibertyClassics, 1987], esp. pp. 16–17).

5. David Hume, "Of Public Credit," in ibid., p. 352.

6. Ibid., p. 350.

7. Ibid., p. 638. These last remarks quoted are from Editions H to P of the *Essays*, as distinguished by Miller, and are from the Variant Readings at the end of the collection of essays.

8. The way we have handled the recent savings and loans crisis is an instance

of this sort of skewing of alternatives. We have put off paying off the debt in such a way that the total cost will be significantly higher than it would be if we were to pay it off more quickly, but that cost will be borne not primarily by this generation of voters, but by the next—and the next.

9. I have in mind here in regard to the nature of a novel the work of M. M. Bakhtin. See *The Dialogic Imagination*, Michael Holquist, ed. (Austin: The University of Texas Press, 1981).

10. Ruth Benedict, *The Chrysanthemum and the Sword* (Boston: Houghton Mifflin, 1946), p. 14. It is the aim of Benedict's book I am trying to emulate, and there is clearly much more to be said about the feasibility of having such an aim, especially given the obvious difficulties involved in having to look at another culture through one's own eyes, in being aware of how that can bias what it is that one perceives, and in the very looking being taken to be, and perhaps being, in some ways exploitative—like a voyeur.

11. Ibid., p. 16.

12. It is thus no surprise to find political analysts remarking, after Clinton's election, on the difficulties he would have in attempting to raise taxes to cut back on the budget deficits since the economy seemed to have turned (see David E. Rosenbaum, "Better Days Muffle Clinton's Call for Sacrifice," *New York Times* [February 7, 1993], p. E1).

Rosenbaum gives a sample of the sort of analysis I am arguing for here when he compares the cast of mind of politicians with that of economists given the good news about the economy. To an economist, he argues, "the time to attack the deficit is when times are good—when tax increases and spending cuts can be weathered without sending the economy into a tailspin." But, he argues, "Politicians see this through a different lens. When the economic picture is bleak, they say, their constituents are ready to do whatever it takes to turn matters around. But when people are back on their feet, it becomes harder to convince them of the need for new unpleasantness, be it higher taxes or lower benefits or cutbacks on favorite programs" (ibid., p. E1).

13. As David R. Mayhew has argued, being re-elected "has to be the *proximate* goal of everyone, the goal that must be achieved over and over if other ends are to be entertained" (*Congress: The Electoral Connection* [New Haven: Yale University Press, 1974], p. 16). If a politician is elected to achieve certain ends, he or she must be re-elected. Few ends can be achieved in one term. To attribute such an interest to politicians is thus not to impugn them. After all, as Mayhew notes, politicians *ought* to be concerned to be re-elected—and so be encouraged to act in ways that further that end—if underlying and justifying the democratic process is that politicians are to be accountable to the public (ibid., p. 17).

The assumption that politicians seek re-election can be construed as a helpful way to stain the political process of the body politic—and I use the metaphor purposefully—so as to highlight those features that are likely to skew both the determination of alternatives and the best choice from among them for environmental issues. Staining it with a different assumption will highlight other features. I do not by any means intend to exclude alternative ways of examining the body politic.

14. It might be added that the *rhetoric* of the political process, in this country at least, is rights, not interests, but in saying that one enters a complex dispute concerning the roles of appeals to rights and appeals to policy in law, among other things. For a brief discussion of *some* of the issues involved, see, for example, Kent Greenawalt, "Policy, Rights, and Judicial Decision," in Marshall Cohen, ed., *Ronald Dworkin and Contemporary Jurisprudence* (Totowa, NJ: Rowman & Allanheld, 1984), pp. 88–118.

15. It may sound in much of what follows as though I am recommending the decision-procedure I am trying to describe. There is a reason for that. Methods of handling problems evolve in response to real problems and usually represent at least prima facie reasonable solutions to those problems. We cannot properly understand a procedure without articulating the reasons that lie behind it. In trying to give those reasons, where appropriate, as well as describe the procedure itself, I may be thought to be arguing for it. But I am not doing that. My aim is the neutral one of laying out its more systemic features so that they can be seen— and so noted if they will make a difference to environmental decisions. If we do not appreciate the reasons that account for dominance of the procedure we are examining, we will be ill-prepared to counter any of its unfortunate consequences.

16. Hume, *Essays*, esp. pp. 513–14.

17. *The Federalist Papers*, Garry Wills, ed. (New York: Bantam Books, 1982), esp. p. 48.

18. It is a political and philosophical question of some import whether an aggregate of interests can outweigh a right and, if so, what right and under what circumstances. It is not one that I need address here because I do not think the interests at issue are so weighty, even in the aggregate, as to justify overriding anyone's rights, of whatever sort. For a discussion of some of the philosophical issues involved here, see Ronald Dworkin, "Hard Cases," in *Taking Rights Seriously* (Cambridge: Harvard University Press, 1978), pp. 82ff.

It is not difficult to imagine someone arguing for the view that aggregate interests override particular rights. For instance, someone who thinks families ought to be limited to one child could readily give such an argument, denying what we presume to be a right to have as many children as we wish on the grounds that the interests of all would be significantly harmed by exercising such a right.

19. Lon Fuller thus says, "It is not so much that adjudicators decide only issues presented by claims of rights or accusations. The point is rather that *whatever* they decide, or *whatever* is submitted to them for decision, tends to be converted into a claim of right or an accusation of fault or guilt" ("The Forms and Limits of Adjudication," 92 *Harvard Law Review* 2 [December 1978], p. 369).

20. The dispute over the incinerator in East Liverpool, Ohio, is a case in point. The residents speak most loudly about the pollution from the plant and the danger to their health, and these interests are represented as rights that would be denied by the incinerator's operation and are being denied by the approval process. The claims they are emphasizing are technical concerns about whether the approval process really was proper. These "technical concerns" are effectively claims that rights to due process in the approval process were denied (see Keith Schneider, "Gore Says Clinton Will Try to Halt Waste Incinerator," *New York Times* [Decem-

ber 12, 1990], pp. B1 and B3). Moving the issue into the judicial system effectively transmogrifies any interests into claims of right, and those will be accepted or rejected depending on how weighty those interests look once given the dress of rights (see Keith Schneider, "Incinerator Trial Is Blocked, Leading to Test of New Administration," *New York Times* [January 18, 1993], p. A15).

Of course, in the case of the Liverpool incinerator, those opposed were correct that rights had been violated—if we have a right that a clear law not be violated. The Environmental Protection Agency admitted in a House Judiciary subcommittee hearing that it had violated the hazardous waste law. It issued an operating permit even though the permit did not contain the names of those who owned the land on which the incinerator was to be built. That is expressly required by law in order "to identify those who are liable in case of an accident." The agency later added the owner's name, but "without any public comment, another violation of the law" (Keith Schneider, "2 Admit E.P.A. Violated Hazardous Waste Law in Issuing Permit," *New York Times* [May 8, 1992], p. A15).

The recent history of the incinerator is itself a case study in the different ways courts and the political process treat rights and interests. The Clinton administration is apparently arguing that because the Bush administration "changed the picture when they granted the test burn permit" in January, the new administration's hands are tied and it must bow to a court decision that sanctioned the incinerator's operation. Now that the courts have decided, it seems to be saying, it has no option. But as the Clinton administration had the authority to rescind the test burn permit on entering office, this looks rather like one political entity passing the buck to another, a concept we examine later, and passing it to the judiciary so that no matter what the decision, it can do nothing because interests cannot properly compete with rights (Keith Schneider, "Ohio Incinerator Cleared over Objection by Gore," *New York Times* [March 18, 1993], p. A20). See p. 194n27.

21. There are exceptions, of course. The Mescalero Apaches have expressed an interest in "storing high-level nuclear waste for at least a few years on [their] reservation in southern New Mexico." The site would "hold spent nuclear fuel until a permanent repository is built," and the pay-off, it is claimed, could be "long-term independence and prosperity" for the tribe (Matthew L. Wald, "Tribe on Path to Nuclear Waste Site," *New York Times* [August 6, 1993], p. A12).

It is not without some irony that "the Governor of New Mexico, both houses of the Legislature, both Senators and all three Representatives are adamantly opposed" to the Mescalero Apaches having an interim nuclear waste dump on their reservation. Although the Apaches want the dump in their backyard, others in New Mexico do not want it in theirs. Part of what is at issue is what counts as someone's backyard (Matthew L. Wald, "Nuclear Storage Divides Apaches and Neighbors," *New York Times* [November 11, 1993], p. A8). Because the Mescalero Apaches are a sovereign nation, it is not obvious that those in New Mexico can have much say over the matter.

22. Kent E. Portney, *Siting Hazardous Waste Treatment Facilities: The NIMBY Syndrome* (New York: The Auburn House, 1991), p. 25. For a thorough discussion of some of the issues from a variety of perspectives, see also Charles E. Davis and

James P. Lester, eds., *Dimensions of Hazardous Waste Politics and Policy* (Westport, CT: Greenwood Press, 1988).

23. Portney, *Siting Hazardous Work Treatment Facilities*, p. 25.

24. Ibid., pp. 138–139. I should say that it is not obvious that the aim is to reduce risks. Portney seems to switch from "perceived risks" to "risks" throughout the text. What Portney *calls* "risk substitution" is introduced as a way of making changes "without worsening individuals' *perceptions* of the risks they face." He does not call it "risk perception substitution," although it is defined that way. Yet, in defining its variants, he speaks of diminishing "existing risks." It is unclear whether the aim is to diminish the perceptions, the actual risks, or the former by diminishing the latter. Because we can alter the perception of risk without lowering any risks at all, and because it is easy to understand why those who have already been put at risk through no fault of their own might think that, once again, the system will work against them, the switching back and forth is not idle.

This alternation of risk reduction with the reduction of a perception of risk pervades the text. For instance, in introducing the economic theory of compensation that he thinks has been ineffective, he says that those near an incinerator "are being asked to bear high personal costs (in the form of risks)"—*not* "in the form of *perceptions* of risk"—so that "the focus has become one of attempting to devise methods of altering people's *subjective* assessments of benefits relative to risks associated with living near potentially noxious facilities" (ibid., p. 25, italics mine).

25. Ibid., p. 139.

26. Mayhew, *Congress*, p. 47. Mayhew is arguing that it is a mistake to suppose that those in Congress act to *maximize* their chances of re-election because they "act in an environment of high uncertainty." "Behavior of an innovative sort," he says, "can yield vote gains, but it can also bring disaster" (ibid.). They are thus more inclined to conservative strategies and, I suggest, tend not to consider what need not be considered.

27. It was only after the smog became "so obvious and odious to the public [as to produce "sudden 'gas attacks' that irritated the eyes"] that elected leaders were compelled to take meaningful action" (James M. Lents and William J. Kelly, "Clearing the Air in Los Angeles," *Scientific American* [October 1993], p. 33).

28. Al Gore has said, while a Senator, "The problem in organizing our response is that the worse effects seem far off in the future . . . while right now, in the present, millions of people are suffering in poverty and dying of starvation, warfare, and preventable disease." His point is that our political awareness and our responses are shaped by these events, pulling us and our ideas as we become aware "of a downslope toward a future" we do not want to see (*Forum on Global Change and Our Common Future*, National Academy of Sciences, Washington, D.C., May 1, 1989, p. 3). See also in this regard Al Gore, *Earth in the Balance* (Boston: Houghton Mifflin, 1992), pp. 2ff.

29. William Orphuls and A. Stephen Boyan, Jr., *Ecology and the Politics of Scarcity Revisited* (New York: Freeman, 1992), p. 50.

30. A quotation from Dr. David Pimentel, "a professor of entomology at Cor-

nell University" ("An 'Unloved' Grass May Turn Hero in the Struggle against Soil Erosion," *New York Times* [February 16, 1993], p. C4).

31. Ibid. I say "something like" because vetiver will only grow in the southwest, and some variant needs to be developed for the Midwest and other parts of the United States.

32. Finding such solutions can be difficult, but that is why politics is an art. Consider Madison's argument in *Federalist Paper* Number 54 as to why a slave should be counted as three-fifths of a person. He there argues that it is in the interests of both North and South to so count slaves, and he ends by noting that because representation in Congress is determined by the population of the various states and because any census will "necessarily depend in a considerable degree on the disposition, if not the cooperation of the States, it is of great importance that the States should feel as little bias as possible to swell or to reduce the amount of their numbers." But if both their representation in Congress and their liability to taxation are determined by population, "the States will have opposite interests, which will controul and ballance each other; and produce the requisite impartiality" (*The Federalist Papers*, p. 279).

Think of interests as lines of force, pulling and pushing persons and political entities one way and another. Madison argues that forces pulling in the opposite direction can be harnessed to a single end and so balance each other. His argument that the North and the South should both support counting a slave as three-fifths of a person is of this sort. The North has competing interests in a smaller number of representatives from the South and a higher proportion of taxes from the South, when taxation is determined by the number of persons, and one way to reconcile those interests, which pull in opposite directions, is to count slaves, but not as full persons. The same sort of resolution holds for the South, which wants to increase its number of representatives but diminish its taxes.

The situation we are examining in regard to pollution concerns an immediate *disinterest* whose drag on action must be neutralized by a solution that appeals both to immediate interests and is also in the long-term interest. For politics conceived as a physics of interests, the problem is to find Madison-like solutions to environmental issues.

33. The explanation for this is considerably more complicated than anything I have suggested here may imply. As Mayhew points out, "satisfaction of electoral needs requires remarkably little zero-sum conflict among members" (Mayhew, *Congress*, p. 82). If we construe the participants' primary interest to be re-election, Mayhew is suggesting that the procedure for resolving competing interests is so designed that it does not require any participant giving up that interest in re-election. So some self-interested behavior is consistent with the wider concern mandated by a consideration of competing interests.

34. It is arguable that some professions do attempt such institutionalization—the ministry, or social work, perhaps.

35. I am, again, not suggesting that considerations other than immediate self-interest do not enter into political calculations. They often do. In the Constitutional Convention, Mr. Gerry suggested that it was "necessary to limit the number of

new States to be admitted into the Union, in such a manner, that they should never be able to outnumber the Atlantic States." His concern was that the original states ought not to put themselves in the hands of the men in those states because they "will if they acquire power like all men, abuse it." Mr. Sherman responded by saying, "We are providing for our posterity, for our children & our grand Children, who would be as likely to be citizens of new Western States, as of the old States. On this consideration alone," he added, "we ought to make no discrimination as was proposed" (*Notes of Debates in the Federal Convention of 1787 Reported by James Madison* [Athens, Ohio: Ohio University Press, 1966], p. 288).

36. This assumes, of course, that those drinking too much choose to do so and thus could take into account the consequences of that choice, short or long term.

37. Hume, "Of Public Credit," in *Essays*, p. 638.

38. Matthew L. Wald, "Designing Tankers to Minimize Oil Spills," *New York Times* (June 17, 1990), p. E5. See also Thomas D. Hopkins, "Oil Spill Reduction and Costs of Ship Design Regulation," *Contemporary Policy Issues*, Vol. X, No. 3 (July 1992), pp. 59–70. It is a distressing feature of discussions of tanker design that these are all carried out only in terms of a cost–benefit analysis, as though we could wring from the complexities of the issue, without remainder, a single value in terms of which we can determine what we ought to do.

39. It is thus no surprise at all to hear, for instance, that a barge "struck a coral reef off the coast" of Puerto Rico and "began spilling its pungent cargo [of at least 750,000 gallons of heavy oil] along the city's premier resort beaches" ("Big Oil Spill Off Puerto Rico Fouls Beach at Height of Tourist Season," *New York Times* [January 8, 1994], p. A1). The barge has a single hull, but the current law allows a 20-year period for retrofitting with a double hull, and "a company spokesman . . . said that the *Berman*, an old vessel, would have been scrapped instead of upgraded" (Ronald Smothers, "Stockpiled Equipment Aids Cleanup of Oil in San Juan," *New York Times* [January 9, 1994], p. A12). We can expect many more such accidents to occur for some time.

The concern not to harm current interests by requiring significant changes now in how tankers and such barges are built has the effect of ensuring that other interests will be harmed. There are no free lunches. The Puerto Rican tourist industry takes in about $2.5 billion, and the spill had adverse economic effects (Ronald Smothers, "After Oil Spill, Puerto Rico Tries to Salvage the Season," *New York Times* [January 11, 1994], p. A18). Of course, one variable is that the current interests being harmed are clearly identifiable whereas the specific interests that will be harmed are not, and that is a variable that makes an enormous political difference. That is, a *thorough* cost–benefit analysis, one that took into account all the economic damage caused by spills, could well provide a justification for radically altering existing tanker designs—without the need for considering any of the other values a cost–benefit analysis wrings out of the situation.

40. Orphuls and Boyan, *Ecology*, p. 221. I will not pursue this line of thought because a proper examination would require another book, but it is a major thread of support for the claims I am making about how the decision-procedure of our political system tends to choose those options in which the benefits are front-

loaded, somewhat independently of the long-term benefits. It is not just that decisions in business seem made in the same way but that the same cast of mind is at work, each reinforcing the other so that it becomes hardly noticeable that it is a cast of mind, an artifact that can be changed.

41. Even such a relatively good example of environmental legislation as the Clean Air Act of 1990 illustrates how difficult it is to choose alternatives in which burdens are up front, despite significant benefits down the road. The general aim of the Act's provisions regarding the selling of rights to utilities to pollute is to allow market forces to "encourage utilities that can cut pollution for the least investment in scrubbers or fuel changes to lead the way." But one of the effects of that aim is to encourage utilities that can to decrease their immediate expenses even if the long-term burdens on the general public are increased. The Illinois Power Company, to pick one example, has "stopped construction on a $350 million scrubber and begun in private deals to stockpile permits that it will use between 1995 and 2000." A spokesman claims that it will save "at least $250 million over the next 20 years and save Illinois coal-mining jobs as well" (Barnaby J. Feder, "Sold: $21 Million of Air Pollution," *New York Times* [March 30, 1993], pp. D1 and D22). So one way of conceiving of the provision of the Act regarding pollution rights is that it harnesses market forces to encourage just the sort of burden shifting and short-sightedness that the political decision-procedure itself encourages. Pollution will not be decreased for those downwind of the Illinois Power Company plants. The burden of getting rid of the pollution has been shifted from the Power Company to those who will breathe it in and from the present to sometime in the future. In short, the bias is evident in that Act where various compromises were made to lessen immediate harms even though the long-term benefits were lessened as well. For a brief description of the Act, see Keith Schneider, "Lawmakers Reach an Accord On Reduction of Air Pollution," *New York Times* (October 23, 1990), pp. A1 and A18.

42. See "E.P.A. Drops Plan to Require Waste Incinerators to Recycle," *New York Times* (December 21, 1990), p. A33.

43. Politicians often have more than their constituents to answer to. New Jersey's decision to put a hazardous-waste incinerator on the Arthur Kill not only has local officials and residents angry, but has met resistance from New York City, "just across the brackish waterway" (Joseph F. Sullivan, "Debate Rages over Site Proposed for Incinerator," *New York Times* [March 28, 1992], p. 28).

The main argument for siting the incinerator may perhaps strike some as an odd one. The side in favor contends that the incinerator meets all appropriate standards, and the sides opposed argue that it will further pollute an already heavily polluted area. This sort of confrontation is standard. But those in favor of the site also argue that *because* the current site is already polluted, that is the crucial argument in its favor. It is better, they claim, to put "an incinerator in an already polluted area . . . [than to locate] it in a sparsely populated area," better, as the rhetoric has it, that it go in "brown fields" than in "green fields" (ibid., p. 28). The political payoff depends on the judgment that those who live in an already polluted area have less to complain about than those whose pristine environment would be harmed by an incineration plant. I would suspect that those in that environment would find this judgment inappropriate, to put it mildly.

The proponents of siting the incinerator where there is already pollution could have given the sort of argument Kent E. Portney suggests that we examined, namely, that those who are there could be convinced to substitute the incinerator's pollution for their current pollution (*Siting Hazardous Waste Treatment Facilities*, p. 142). But those who live there would still object, I should think, that they would not have been offered such a trade if they did not already have "brown fields," even if through no fault of their own, and that they should not have to accept the substitution, unless, of course, they were to be left with brown fields unless they did.

44. Joseph P. Fried, "Trash-Transfer Plan Stirs Up Queens," *New York Times* (September 6, 1991), p. B3.

45. A harried employee of a school system was asked in a television interview why he had not anticipated the problems with asbestos. He said, "Three years ago I didn't even know how to spell it."

46. Mary Douglas and Aaron Wildavsky, *Risk and Culture*, (Berkeley: University of California Press, 1983), pp. 49–50.

47. "The expert opinions that comprise the scientific literature are typically considered to be 'objective' in two senses, neither of which can ever be achieved absolutely and neither of which is the exclusive province of technical [that is, scientific] experts. One meaning of objectivity is reproducibility: one expert should be able to repeat another's study, review another's protocol, reanalyze another's data, or recap another's literature summary and reach the same conclusions about the size of an effect. . . . The second sense of 'objectivity' means immunity to any influence by value considerations. One's interpretations of data should not be biased by one's political views or pecuniary interests" ("'The Public' vs. 'The Experts': Perceived vs. Actual Disagreements About Risks of Nuclear Power," in *The Analysis of Actual Versus Perceived Risks*, [New York: Plenum, 1983], p. 237). In suggesting that scientists are less than objective in making recommendations about public policy matters, I am suggesting that values other than money and those embodied in the choice of a political party are of importance, especially those regarding how to respond to risk.

48. What would be really depressing would be if the change in stance I am suggesting were made and the complex natural social artifact that is our political system were to accommodate itself to that change, making whatever adjustments are necessary to produce the same sorts of results we now tend to get.

49. Looked at from the point of view of companies, such a problem presents difficulties of economies of scale. If a company manufacturing a product must meet different standards in different municipalities or even different states, the cost of manufacturing soars. That is one reason companies are asking for national standards for what counts as "recyclable." Rhode Island requires that no product be marked "recyclable" unless collection systems and recycling plants are in place for 75 percent of the population. By such a standard only aluminum cans and plastic soda bottles would be "recyclable." In New York, a package can be

labelled "re-usable" only if it can be re-used a minimum of five times, and "recyclable" can be used on a product only if the product contains material that had been previously used. Clearly, any standard chosen will create problems for companies, but having some national standard will alleviate the difficulties caused by "a patchwork of local laws [that] would destroy economies of scale and lead to higher prices" ("'Recyclable' Claims Are Debated," *New York Times* [January 8, 1991], p. D5).

50. The city of Victoria, Canada, every day dumps "20 million gallons of raw, untreated household sewage . . . into waters shared by Canada and the United States" (Timothy Egan, "A Deluge from Canada of Raw Sewage and Autos," *New York Times* [May 19, 1991], p. A1). There is nothing cities in the United States south of Victoria can do except protest and ask its citizens to boycott Victoria. Victoria officials downplay the problem. Analysis has shown that "toxic waste and chemicals, in addition to toilet paper, feces, and bacteria, are sent out to sea." But the chief engineer of the agency that controls Victoria's sewage, Michael Williams, says, "The sewage is 99 percent water, and the other 1 percent is made up of waste from us humans and chemicals we throw away. The concern that we have at the technical level is chemicals. With respect to human waste, mother nature tends to take care of that." But even though there is "concern" about "the technical level [of] chemicals," "Victoria officials say there is no proof that the sewage is harmful to marine life" (ibid., p. A14). It should be no cause for surprise for anyone familiar with how arguments generally run regarding environmental issues that a municipality that must pay up to $350 million to clean up its sewage for the primary benefit of the inhabitants of another country would play the epistemological card. The natural consequence to draw from this issue is that what is needed is a treaty between the United States and Canada, but, in fact, that may not be so. As John Cashore, British Columbia's Minister of Environment, said, "The province has made it clear that dumping of raw sewage is unacceptable. The municipality knows the handwriting is on the wall" (Clyde H. Farnsworth, "Such a Lovely City! Still, Seattle Holds Its Nose," *New York Times* [April 20, 1993], p. A4).

51. One consequence of the cost differential is that freighters are carrying such cargoes from Europe to other sites around the world, and owners are tempted to "dump poisonous chemicals in the ocean to avoid the risk of being stuck with a cargo that they could not land in any country" or in such countries as Nigeria. The *Karin B* was legitimately chartered by the Italian government to remove 3800 tons of chemical waste "clandestinely dumped in an open field . . . in Nigeria." Nigeria "held an Italian freighter and its crew hostage and threatened to break off diplomatic relations" to get Italy to remove the waste. The Italian government was unable to bring the material to Italy because of local protests, was refused permission to unload its cargo in Great Britain, and ended up bringing the freighter back to Italy where there were further protests (Barry James, "Tramp Freighters Ply High Seas With Dangerous Cargoes," *International Herald-Tribune* [September 2, 1988], p. 1). The travails of the *Karin B* are a living metaphor, like the barge *Mobro*, of the convoluted workings of our waste disposal system.

52. Ferdinand Protzman, "A Nation's Recycling Law Puts Businesses on the Spot," *New York Times* (July 12, 1992), p. F5.

53. The sort of law Germany has adopted may have its problems. The costs of collecting everything are enormous, for instance. But one effect such a law will have is that it will encourage businesses to consider, in packaging their products, how best to reduce the cost of collection and disposal. One effect of the law is to encourage significant recycling because, with the right packaging materials, that presents a more efficient way of getting rid of the waste than landfilling or incineration. So glass and aluminum tend to win out over plastic (ibid.).

54. Keith Schneider, "For Communities, Knowledge of Polluters Is Power," *New York Times* (March 24, 1991), p. E5.

55. Richard W. Stevenson, "Monitoring Pollution at Its Source," *New York Times* (April 8, 1992), p. D1.

56. Ibid., p. D1. As I have noted, the idea of a market in pollution is not without its problems. As we saw, the Clean Air Act of 1990 created the right to sell and buy pollution credits, but those east of the Midwest fear "that the biggest purchasers of pollution credits will be the Midwestern coal-burning utility plants that are among the chief sources of acid rain in the East" (James Dao, "A New, Unregulated Market: Selling the Right to Pollute," *New York Times* [February 6, 1993], p. 1). Those downwind of a plant that pollutes will not benefit at all from the plant's buying pollution credits. As John F. Sheehan, a spokesman for the Adirondack Council, an enviromental group, puts it, "The trading program didn't take into consideration where the pollution would fall after it was traded. . . . [So] We have Lilco [a plant on Long Island], whose pollution will go out to sea, trading to the Midwest, where the pollution will fall on us" (p. 1). Lilco can hardly be faulted for selling the rights, however, because it had already, in response to a New York State law, reduced its emissions below the federally mandated level and "has 40,000 to 50,000 allowances a year to sell," with each allowance worth from "$250 to $400" (Matthew L. Wald, "Lilco's Emissions Sale Spurs Acid Rain Concerns," *New York Times* [March 18, 1993], p. B2).

In short, unleashing market forces on pollution has created an enormous problem of trying to predict where pollution will be decreased and where it will be increased. If one aim of the Act was to decrease pollution for the Adirondacks, for instance, then that aim seems at odds with the aim of encouraging market forces to decrease the total "cost of meeting pollution-reduction goals" (Feder, "Sold: $21 Million of Air Pollution," p. D1). As can only be expected for any market, new players have entered. Amex Energy Inc. of Greenwich, Connecticut, has purchased pollution rights for resale, and the worry is that it will sell the rights to a utility whose pollution will continue to harm the Adirondacks, for instance. As John Sheehan, put it, "Part of the problem with this process is there is an inherent lack of accountability as far as what will result from the trading of allowances" (Wald, "Lilco's Emissions Sale Spurs Acid Rain Concerns," p. B2).

57. The value of keeping track of the hazardous output of factories and so on is illustrated by the discovery of a scheme to purchase "waste oil from service sta-

tions and tank cleaning companies in seven states . . . and in Canada" and then mix the contaminated oil with clean oil and sell it as fuel "through independent contractors." The scheme was uncovered "because of a feature of New Jersey law that treats waste oil as hazardous waste and requires the movement of the oil to be documented," and when something was discovered "wrong with some manifests," the ensuing investigation led to that scheme and to another where No. 2 oil was purchased from refineries tax free and then sold "to filling stations for sale as diesel fuel," "thereby avoiding millions of dollars in Federal and state taxes" (Joseph F. Sullivan, "12 Held in Trucking of Untaxed and Contaminated Oil," *New York Times* [May 28, 1993], p. B5).

58. Alan R. Gold, "Ripples from a 1988 Rule Led to Messy Trash Strike," *New York Times* (December 12, 1990), p. B1.

59. The State of Ohio agreed to allow the Southern Ohio Coal Company to dump "a billion gallons of polluted water into 75 miles of streams and creeks that flow into the Ohio River." The water is in a flooded coal mine, and if it is not pumped out, the mine will close, costing "815 jobs, or $78 million a year in lost wages and benefits in a corner of the state where a quarter of the population lives below the poverty level," and costing "$34 million worth of mining equipment." The argument for dumping is that "the environmental damage is temporary and reversible," but federal officials tried to stop it, saying that "the company was merely using the politically powerful issue of jobs to avoid the expense of purifying its waste water" (Michael deCourcy Hinds, "Coal Mine Sends Polluted Water into Ohio Creeks," *New York Times* [August 20, 1993], p. A12). It is an odd situation where a determination of what is best for the environment and for the economy must be a matter of dispute between local and federal officials and turn on a court decision about whether the federal authorities have jurisdiction rather than on any determination of the merits of the case. I say this independently of any judgment of the merits of either side to this dispute.

60. Harvey Lieber, "Federalism and Hazardous Waste Policy," in James P. Lester and Ann O'M. Bowman, eds., *The Politics of Hazardous Waste Management* (Durham, N.C.: Duke Press Policy Studies, 1983), p. 60. As it is always a question whether a local authority has a right to act, any attempted act will be met by legal action by those opposed to it. So local action will thus be tempered by the realization that a legal fight will ensue before anything can be done.

5. The Options

1. *Environmental Quality*, 21st Annual Report of The Council on Environmental Quality (Washington, D.C.: Superintendent of Documents, 1990), p. 361.

2. "E.P.A. Drops Plan to Require Waste Incinerators to Recycle," *New York Times* (December 21, 1990), p. A33.

3. William F. Schmidt, "Trying to Solve the Side Effects Of Converting Trash to Energy," *New York Times* (May 27, 1990), p. E5.

4. John H. Cushman, Jr., "A Battle with Echoes: Floridians Fight Power Line," *New York Times* (January 28, 1994), p. A12.

5. See for example, Allan R. Gold, "New York City Is Proposing Tighter Water-Quality Rules," *New York Times* (September 12, 1990), p. B3.

6. Perhaps it needs to be emphasized again that when I argue that the array of options available for any political choice are skewed by the pressures for those options whose benefits are, for example, front-loaded, I do not mean to suggest that such options are thereby necessarily mistaken. It may be that landfilling was the best choice, all things considered, when such choices had to be made. That would be a coincidence, but a happy one, and nothing I have said is meant to suggest that such coincidences are not possible. Of course, such coincidences are no doubt rare. As we will see when we look at the Stringfellow hazardous waste site, making decisions about what is safe and what is not is extremely difficult with the kind of epistemic shortfall typical of waste management.

7. As the chief engineer at Fresh Kills, Michael J. Massi, put it, "When I started here in 1973, there was at least one landfill in each borough. This is the last show in town" (Matthew L. Wald, "Buying Time by Buying Space on Mount Garbage," *New York Times* [June 11, 1993], p. B4). How long that show will last is a nice question, one subject to such contingencies as whether, by the year 2000, "the city recycles or composts half of residential, institutional, and commercial waste *and* can operate incinerators that consume 3,750 tons a day . . . *and* private haulers continue to haul half the city's commercial waste out of the city." The city would then buy anywhere from 28 to 37 years, depending on how much a cubic yard weighs in trash. Fresh Kills's limit is stated in cubic yards, and trash is weighed in pounds, but the density of trash may come to vary in the future if, for instance, recycling and incineration take off the heaviest parts of it (ibid., p. B4, and pp. B1 and B4 for a thorough description of Fresh Kills, its capacity, its currently projected height, and the problems associated with trying to fill it as full as possible).

Although those in New York City responsible for the disposal of its waste are hardly county commissioners, the problems they face are the same. It is a measure of how much incineration is driving the whole that without incineration as a major component, it is argued, "The giant Fresh Kills landfill in Staten Island may have 10 more years before its capacity is exhausted" (Allen R. Gold, "Garbage Can't Just Go Anywhere, Anymore," *New York Times* [December 16, 1990], p. E16). But "New York residents and businesses generate at least 28,00 tons of garbage daily" (ibid.), the amount is increasing, and what comes as a given must be handled somehow. We can imagine the problems facing the political bodies charged with getting rid of waste all over the country by asking where in New York City another landfill the size of Fresh Kills could be placed. Creating another landfill that size would only give the City about 20 to 30 years breathing space at best—if that phrase is not an oxymoron. So we must ask where the next one could go after that, and so on and so on. Fresh Kills is "the last show in town."

8. The Stringfellow hazardous waste site in Glen Avon, California, is typical of the sort of problems local and state authorities find coming back to haunt them. Howard L. Halm is a lawyer defending the State of California in a suit in-

volving about 4,000 plaintiffs regarding the landfill in which "34 million gallons of chemicals [were] dumped in an unlined 20-acre canyon," including "suphuric and hydrochloric acid; volatile organic compounds, including methylene chloride used in paint strippers; a number of pesticides, many containing D.D.T.; caustic chemicals like sodium hydroxide; solvents like trichloroethylene, which is used in dry cleaning; and heavy metals, including lead, manganese, nickel, cadmium and chromium." Mr. Halm argues that the public "had ample input" into the original decision in the 1950s and that, in any event, the State is not liable for any problems with the dump because it only issued a "land-use permit," like a license, and is not "liable for the conduct of a person to whom that license or permit is issued." The State was held liable in a federal-court decision for 85 percent of the damages for cleaning up the site, estimated to "run as high as $800 million" more than the $150 million already spent, and it was held "negligent in choosing, designing and supervising the Stringfellow dump." California is appealing that decision, and its lawyers have "warned that California could be 'exposed to astronomical liability' if appeals fail" (Nick Madigan, "Largest-Ever Toxic-Waste Suit Opens in California," *New York Times* [February 5, 1993], p. B16). Such a suit is every political authority's nightmare, and it is under the pressure of such possibilities, among other things, that landfilling is no longer viewed as the obvious solution.

It is worth remarking that the plaintiffs in such a suit are at a remarkably ironic disadvantage. When original decisions were made about siting such dumps, those opposed were required to *prove* that damage *would* occur from them—a proposition difficult to sustain when the epistemological card can be played at any time and when the sorts of chemicals to be dumped, such as PBB, for instance, have never been tested for their effects on humans and when no one knew ahead of time exactly what would be dumped. Now those who feel that damage has occurred are finding it "difficult to hold any person or agency responsible for their ills." It is not just that it is difficult to determine who exactly is responsible but that, even if they could determinate a liable party, they would have " 'to prove what got to where and how it affected whom'," as Thomas A. Kearney, another of the state's lawyers put it. " 'That's a very tough burden to sustain' " (ibid.). The framework of decision making within which such decisions must be made requires that the standards be set so high that those who are concerned about harm are hurt coming and going. It would be a hard task, and inappropriate, to lower the threshold for showing harm in the judicial process, but it ought not to be so hard to change the framework of decision making so that those opposed to potential harm are not so heavily burdened.

9. After dumping at sea was prohibited, Nassau County agreed to build two plants to turn the sludge produced by treating raw sewage into fertilizer pellets. But local opposition to the plant sites led the county to contract with a company to ship the sludge to West Virginia (Jonathan Rabinovitz, "West Virginia Eyed in Sludge Disposal," *New York Times* [March 2, 1993], p. A15). The plants would have been "within 1,500 feet of two elementary schools" and "within hundreds of feet of homes in Bay Park" (Jonathan Rabinovitz, "U.S. Judge Lets Nassau Sludge Be Shipped Out, Not Shaped," *New York Times* [March 20, 1993], p. A26). It is not an

accident that Richard M. Kessel, "the executive director of the New York State Consumer Protection Board," who is expected to "run for county executive next year," claims that shipping the sludge out of Nassau County will protect "public health and safety, while at the same time" save "taxpayers millions of dollars" (ibid.).

The Nassau County example is paradigmatic of how local political authorities are under increasing pressure to solve the trash problems they face without paying political prices they, as local politicians, can ill afford to pay. The decision may be beneficial to the residents of Nassau County, but it makes it difficult for federal officials who had argued against Nassau County's plan on the grounds that it and the precedent it establishes create "long-term uncertainty" (ibid.). So the Nassau County solution is also an example of how a general problem is being resolved at the local level, with all the attendant difficulties for national solutions. Presumably not every county can do what Nassau has done. Many could not afford the $250 million for the 25 years Nassau has contracted for, and West Virginia, or any other place, is not likely to be willing to accept more and more sludge (ibid.).

10. Raising fees can have even more dramatic effects as we have noted in a different context in Chapter 3. New York City more than doubled its fees at the Fresh Kills landfill to help prolong its life, and the effect was that those dumping at Fresh Kills went as far away as Ohio and Pennsylvania to find cheaper dump fees (Allan R. Gold, "Ripples from a 1988 Rule Led to Messy Trash Strike," *New York Times* [December 12, 1990], p. B1). Of course, as those sites fill up, the problem will arise again.

Illegal dumping is always a problem, but one that increases as dumping fees increase. It is an easy way to increase profits if a hauler picks up trash and then just dumps it somewhere. It is an example of how pervasive the problem is that in New York City last year "the Sanitation Department removed at least 133,000 tons of refuse dumped in city-owned lots," and that total does not count the amount not removed from other lots, both city owned and privately owned (Michel Marriott, "Mere Men vs. Mountains," *New York Times* [August 15, 1993], A37).

There is little a local municipality can do about such dumping except police their vacant lots, and such dumping is bound to increase unless some other form of accountability for waste begins to dominate. Making manufacturers responsible for the disposal of their products once their useful life is past and for the material used in wrapping and shipping their products would be one way to change the calculus so that dumping no longer is as profitable as, say, returning the packaging and used products to their manufacturers. The principle is the simple one that the polluter pays (Joanna D. Underwood and Bette Fishbein, "Making Wasteful Packaging Extinct," *New York Times* [April 4, 1993], p. F13). Such a principle would also be "a powerful incentive to rethink design, packaging and marketing decisions" (ibid.).

11. The road signs in Allenwood, Pennsylvania, I refer to at the beginning of Chapter 2 make much of the effect of an incinerator on the valley and on the perception of others of the valley—"Farmland, not Wasteland." The same point holds for any place taking on the waste of others.

12. It is a sign of how volatile the trash business is that New York City is con-

sidering shipping trash to out-of-state landfills and incinerators because fees have dropped as owners with landfills that face closing under pressure from new governmental regulations "try to fill [them] before closing them" and as incinerators find themselves strapped for trash to burn. The issue is complicated because the city wants a long-term commitment, and no one can guarantee that regulations will not change the economic incentives that currently make shipping out of the city economically viable (Matthew L. Wald, "New York City Considers Exporting Trash," *New York Times* [September 27, 1993], p. B4).

13. "Recycling adds costs by forcing trash haulers to add specialized trucks and extra crews to handle materials separately. And municipalities, or the private companies they contract with, have to build and staff facilities to separate recyclables from one another, and from common trash." The market value of the recycled goods left at curbsides averages anywhere from $30 to $50 a ton, according to different studies, whereas the cost of collection ranges, by those same studies, from $50 to $175 a ton (John Holusha, "Who Foots the Bill for Recycling?," *New York Times* [April 25, 1993], p. F5).

14. The problems created by recycling are enormous. A governmental entity that commits itself to recycling must "process and market the tons of paper, glass, metal and plastic" its sanitation department collects (Allan R. Gold, "As Trash Is Recycled, Where Can It All Go?," *New York Times* [October 3, 1990], p. B4). It is some measure of the difficulty governmental entities face that New York City's recycling law "anticipates collection of 3,400 tons a day" in 1993 while its first "major city-owned processing center, a 300-ton-a-day operation planned for Staten Island, is not planned to be completed until April 1993" (ibid.).

A governmental entity committed to recycling also has to find markets for what it recycles, but finding markets is not easy. The City of New York must *pay* "paper dealers as much as $25 a ton to take the material because the market is glutted" (ibid.). In fact, one of the unintended consequences of the efforts to recycle newspaper and magazines in the United States is that "it's cheaper to haul surplus paper all the way to Europe rather than storing or burning it in America," as Paul Nouwen of the Netherlands Environmental Ministry said. He added, "It's offered here practically for free and [has] ruined the market in all of Western Europe. In Holland, the price has dropped from 8 cents to 1 cent or less per kilo just this year" (Marlise Simons, "U.S. Wastepaper Burdening Dutch," *New York Times* [December 11, 1990], p. A7). The Dutch, in particular, have been recycling paper for over fifty years, and whole industries have grown up to handle it. The glut of paper from the United States is undermining the entire system, driving more than 4500 people out of business (ibid.). One other perhaps more paradoxical consequence is that because the unused paper costs money to store, it may end up in landfills.

The lesson of what is happening to waste paper is worth drawing out. Recycling is driven in part by the economic benefits of selling what can be saved, but an additional paradox of recycling is that as more and more of a particular product such as paper is saved, the market becomes more glutted, prices fall, and there is thus less incentive to recycle (Keith Schneider, "As Recycling Becomes a Growth Industry, Its Paradoxes Also Multiply," *New York Times* [January 20, 1991], p. E6).

These problems may just be temporary growing pains, but they are nonetheless real problems for those, like the Dutch, faced with mounting piles of paper and collapsing paper collection systems.

Against such a background, it is of real importance that a commitment be made by heavy users of paper to opt for recycled paper when it becomes available. As John Ruston, of the Environmental Defense Fund has remarked, "The paper industry is the most capital-intensive one in the United States. They are not going to invest in new technology without the assurance of demand." It is with the idea of the need for a prior commitment that the Environmental Defense Fund sought out heavy users of paper in a variety of fields to get a commitment from them to use recycled paper. Six big companies were approached and made that commitment—from companies, Ruston said, providing financial services such as Prudential and Nationsbank, to "packagers with Johnson & Johnson and McDonald's, and a full-service publisher with Time, including book publishing" (John Hulusha, "An Alliance of 6 Big Consumers Vows to Use More Recycled Paper," *New York Times* [August 19, 1993], pp. A1 and D12).

The basic point is that because recycling is "at its heart . . . an industrial process," no one is going to invest money in turning what is recycled into a product unless there is money to be made by selling that product. Ensuring customers for the product ahead of time provides the incentive needed for investment. "When a company that produces one billion magazines a year, as Time does, says it wants to buy recycled paper, that represents a business opportunity for the paper mill that installs recycling equipment and a risk for those that do not" (John Holusha, "Environmentalists Try to Move the Markets," *New York Times* [August 22, 1993], p. E5).

15. It is certainly arguable that recycling is not now profitable—especially if we do not take into account any other factors such as the costs saved through, for instance, not having to landfill what is recycled. But we should keep in mind that a thoroughgoing program of recycling, once in place, can garner great savings. The experience of Terre in Liege, Belgium, is a good example of how a commitment to recycling can pay off. Terre collects "450 tons of used clothes, suitable for resale at home and abroad," every month from Belgian households as well as "65 tons of paper and cardboard," some of which is recycled "into insulation panels." It makes $10 million a year, paying nothing but collection costs for its raw material (Marlise Simons, "Gold in Streets (Some Call It Trash)," *New York Times* [January 4, 1994], p. A4). What it collects is kept out of the waste municipalities have to handle somehow, and its well-working procedures are a model of how recycling can work effectively, with great economic efficiency.

16. The reasons are straightforward. Any attempt to build an incineration plant will be met with resistance by those within proximity of the plant, and it is difficult to know, before building the plant, how much the political and legal battles will cost and whether they will be won or lost. If they are lost, one may lose the money invested in the plant as well as the costs of legal defense. The same sort of story can be told regarding landfills and recycling centers. This source of uncertainty in projections of cost is only one of many. For instance, as more and more communi-

ties become concerned about lead in their water supplies, more and more tests of ground water for contamination and for *possible* contamination become necessary. In short, changing conditions such as increased awareness of and concern about lead poisoning create new problems that cannot be foreseen when calculations are made comparing costs of the various methods of disposal for a community.

17. I again have in mind here such simple examples as trying to purchase milk in pouches that can be readily recycled and produced using far fewer resources.

18. Linda Greenhouse, "Justices to Decide If Ash Is a Hazardous Waste," *New York Times* (June 22, 1993), p. A19.

19. The track record is a relatively short one. We are only now coming to grips with the problems produced by landfills that passed all the criteria for success in the 1950s, for instance. There is no guarantee that forty years from now we will not also find that we have problems with incinerators that, from today's perspective, look problem free. We have not lived through the life cycle of an incinerator. So we may discover that we will face major problems as incinerators age that we cannot now predict or that, in the worst sort of case, we face the same sorts of problems regarding incinerators that we face regarding nuclear plants.

An incinerator presumptively has a limited life span, like any piece of equipment, and it will not only wear out but also become obsolete because of technological advances. How long that span may be is now unknown for sure, and when its end is reached, there will be the problem of what to do with the structure that may be contaminated by the heavy metals that cannot be incinerated. The Shoreham nuclear-power station is an object lesson here. The costs of decontaminating it have been high—over $170 million—and the costs of tearing it down are so high that it has been judged not worthwhile. So it "may simply be dismantled and abandoned, a ghostly monument," leaving 11 acres of prime real estate unusable (Peter Marks, "Shoreham's Last Gasp Brings Issue of Options," *New York Times* [August 20, 1993], p. B5).

20. I would hope that most of the criteria I have given will strike the reader as fairly unproblematic, the sort of considerations that anyone concerned to solve such a problem and to do so efficiently would think of. In fact, some of these criteria are tied to the way we think about waste as something to be managed. For instance, one criterion, that the preferred solution will cost little, depends on the claim that waste is not a product, so that money spent on it is a waste, but that way of thinking about waste is itself an artifact of our practice. If we were to design products so that they or their component parts could be recycled, they, or their parts, would be on their way to becoming products as they went through their own life cycle. We could imagine ourselves wondering whether something had finished its useful life as one thing because we were in need of it, or of some of its components, as something else.

I do not mean to suggest that these are the only criteria a political body such as a county commission would or should consider, only that these criteria are the ones that tend to dominate discussions in such political bodies. The evidence for this comes from many a long night spent at hearings about what such political bodies as county commissions ought to do. That is a limited source, and that is why I must appeal to the reader's sense that these are plausible and relatively unproblematic criteria.

21. David Hume, "Of Public Credit," *Essays, Moral, Political, and Literary*, Eugene F. Miller, ed. (Indianapolis: LibertyClassics, 1987), p. 638.

22. "Trash-to-Electricity Disposal Lags in Maine," *New York Times* (January 22, 1991), p. A19.

23. In projecting costs for incinerators, for instance, any company must take into consideration the possibility that after all the expense of planning and then building a plant, the completed plant either may not be permitted to operate or may be closed after too short a period of operation to recoup much of the costs. If the incineration plant in East Liverpool does not continue operations, the Swiss company that owns it will have a plant that it can neither use nor sell—a dead loss. Its failure "would be the latest in a string of hazardous waste incinerators closed in North Carolina, Kentucky, Illinois and Arizona since 1990. The first was shut because of toxic emissions that harmed workers, two more suffered explosions, and public opposition compelled Arizona Governor Fyfe Symington, a Republican, to buy an incinerator near Phoenix in 1991 before it opened" (Keith Schneider, "The Environmental Fix With a Legion of Doubters," *New York Times* [Dec. 20, 1992], p 132. So there is good evidence in projecting costs for any proposed incinerator that one must take into account the real possibility of not recouping expenses.

To consider another example, Browning-Ferris Industries received about 20 percent less than it had expected for "its state-of-the-art hazardous waste disposal site 70 miles east of Denver" (Barnaby J. Feder, "A Disappointing Deal for Browning," *New York Times* [February 22, 1991], p. D1). The company "intended to completely close the books on its hazardous waste business," but, in addition to selling the plant for significantly less than expected, it had to agree to a provision that tied the ultimate price to the plant's performance (ibid., p. D4). The disappointing sale "highlighted uncertainties peculiar to environmental businesses, especially those in fields where the investment demands—as well as the regulatory and technology risks—are substantial" (ibid., p. D1).

24. We can ask, for instance, whether the "trash-to-energy concept is fundamentally flawed or has simply been poorly executed" and whether two incinerators can ever be more than an experiment because evidence on the operation of many more for a much longer time would be necessary to draw any firm conclusions ("Trash-to-Electricity Disposal Lags in Maine").

25. I do not mean to imply that changing the driving force for waste to reduction, recycling, and reconfiguration for recycling will not itself produce problems. It has been a primary concern of environmentalists that building incinerators produces a need for trash, and one effect of having so many incinerators has been an increased concern by local authorities to control the flow of trash from and into their areas. A similar problem arises if one of the animating principles is recycling, for if local authorities "are to justify a substantial investment in trash-handling, they must be assured of receiving all local garbage and recyclable materials—that without this guaranteed volume, the system does not make economic sense." Recycling has the same effect on trash that incineration does: It turns waste into a

profitable product. The concern is that "the payback for different materials varies widely. If the private sector can drain valuable material out of the system, it could change the economics for local governments, making it harder for them to float loans or pay them off" (John Holusha, "Here's a Switch: Now They're Fighting Over Garbage," *New York Times* [January 23, 1994], p. F8).

26. "There has never been a more potent symbol of the national conflict between convenience and conservation. More than seventeen billion disposable diapers were sold in the United States last year, and every child who uses them goes through about 4,500 in his or her infancy" (Michael Specter, "Among the Baby Set, Disposable Diapers Are Back," *New York Times* [October 23, 1992], p. A1).

"Three years ago 22 states considered taxing or banning disposables. None have succeeded" (ibid.). The efforts of environmentalists have failed to convince parents that the environmental savings are worth the loss of convenience. One difficulty is that it is not obvious that the savings are that significant. It takes water and detergents and energy to wash cloth diapers, and on the West Coast, for instance, where water can be scarce, disposable diapers may be more environmentally friendly than cloth ones. Besides, it is not obvious that disposable diapers take up as much space in landfills as it has been claimed they do. "Excavations of representative landfills—including Fresh Kills on Staten Island—have revealed that discarded diapers take up from 0.5 to 1.8 percent of landfill space" (ibid., p. B2). That is much less than old newspapers, for instance.

27. It is a measure of how intractable a problem our use of diapers presents that first Proctor & Gamble and most recently an independent inventor have tried to recycle disposable diapers. Proctor & Gamble gave up and "instead has sunk $20 million into the development of a composing system that will turn diapers into fertilizer," but the independent inventor thinks he has a method of recycling that will be cost-effective and is efficient (Teresa Riordan, "Patents," *New York Times* [July 26, 1993], p. D2). The advantage of such a method is that it will finesse the difficult choice between cloth and disposable diapers: If disposable diapers can be recycled, they can be preferred without paying a heavy environmental cost.

28. This leaves untouched the one-tenth of 1 percent, and as James J. Donahue, Jr., director of corporate communications at Duracell Inc. put it, "Batteries are by their nature not green. They are little packages of chemicals." But Duracell now produces "mercury free" alkaline batteries—free, to be accurate, of any added mercury although they "may contain small amounts of mercury found in raw manufacturing materials" ("Duracell Acts on Mercury," *New York Times* [March 5, 1992], p. D24).

There is the problem that smaller button batteries require significantly greater quantities of such toxic metals as mercury, silver, cadium, and lithium in order to obtain greater efficiencies (John Holusha, "Trying to Make Batteries 'Green,'" *New York Times* [June 3, 1990], p. F9).

29. John Holusha, "New Packaging That Spares the Environment," *New York Times* (March 28, 1990), p. C6.

30. John Holusha, "New Plastic in Heinz Bottles to Make Recycling Easier," *New York Times* (April 10, 1990), p. C5.

31. "New Packaging That Spares the Environment," p. C6.

32. I have in mind here the containers for fruit juices that are claimed by their manufacturers to be recyclable. These are known in the trade as aseptic packages, and one recent newspaper advertisement claimed, "Drink boxes are as easy to recycle as this page" (John Holusha, "Drink-Box Makers Fighting Back," *New York Times* [December 15, 1990], p. 33).

Another ad lists the virtues, using a check list, that drink boxes are supposed to meet, "the EPA's weapons against solid waste." Drink boxes are supposed to "use 50–75% less material than most conventional packaging"; they "can be recycled," it is claimed, "into everyday items like tissues, envelopes and paper towels"; "the energy generated by burning a pound of empty drink boxes is equivalent to that generated by a pound of coal"; and "the percentage of landfill space consumed by drink boxes is negligible . . . only . . . approximately 3/100 of 1%" (*New York Times* [November 2, 1990], pp. A12–13).

Assessing these claims is complicated. For one thing, it often seems as though apples are being compared to oranges, as when the boxes are claimed to use less material than conventional packaging, and, for another, it often seems that the claims have an other-worldly aspect to them, as when it is claimed that they can be recycled when in fact they are never collected for recycling and few, if any, facilities exist to recycle them.

It is worth noting that (1) to do a proper job of comparing the worth of drink boxes with other forms of packaging, one would have to show that they are environmentally better than any other form of packaging, including pouches, for instance; (2) it must at least be shown that incineration of drink boxes does not produce new pollutants; (3) it would help to show that drink boxes can be collected for recycling economically; (4) the data are somewhat suspect since Tetra Pak, Inc., the largest company that produces the containers, had sales in 1990 of over 60 billion cartons worldwide (see Holusha, "Drink-Box Makers Fighting Back"), making the claimed consumption of landfill space questionable; and (5) the claims that drink boxes use less material than most conventional packaging is suspect given the little material used, say, in pouches, to pick just one example, and the number of layers and kinds of material used in drink boxes, which consist of six layers of polyethylene, paper, and aluminum foil.

6. What Ought We to Do?

1. According to some standard, of course. If we are planting daffodils, one standard would presumably be an aesthetic one. Does this planting or that produce an aesthetically more pleasing display? Determining the appropriate standards in environmental matters is part of the point of this chapter.

2. We also need to consider whether we shall produce new effects by not acting. It is possible that circumstances have so altered that a decision to leave things "well enough alone" will in fact produce different effects than would have before been produced by such inaction. Inaction has its effects too.

3. Many of life's little disagreements seem caused by these sorts of problems.

Two agree to spend a quiet evening together, and one envisages reading to music whereas the other presumes they will be going out. It would not be so bad if such problems were limited to private matters, but they are endemic to public policy issues. It ought to be presumed the rule, rather than the exception, that the ends to be achieved are essentially vague and essentially contestable. For instance, Greenpeace argues that many of the chlorinated organic chemicals should be banned, including the chlorine used to disinfect municipal water supplies, because "so many have turned out to be both toxic and environmentally persistent" (John Holusha, "Greens Pick an Enemy: Chlorine, the Everywhere Element," *New York Times* [December 20, 1992], p. E2).

4. It is much more likely than not that any natural social artifact will be in tension, having ends that compete with each other. It is characteristic of any natural social artifact—from a relationship between two persons to a family to a corporation to a society—that it exists for ends that can come into conflict, and often do, and the form any argument tends to take over such a conflict concerns which end is the *real* end of the artifact. So I do not think the problems we have with our current natural artifact of waste management are peculiar to it.

5. Connecticut recently passed a mandatory recycling law, the third state in the nation after New Jersey and Rhode Island to do so, and one of the complications with making it successful, it is claimed, is its commitment to incineration. Connecticut incinerates 60 percent of its garbage, for instance, more than any other state, and Neil S. Seldman, director of waste utilization for the Institute for Local Self-Reliance in Washington, says that "Connecticut is not doing well in recycling—the state's policy is a quagmire. The state will never get anywhere until it phases out its incineration," whose purpose, it is being claimed, clashes with that of recycling (Nick Ravo, "Few Towns Ready for Connecticut Recycling Law," *New York Times* [January 22, 1991], p. B1).

6. The phrase is from Gordon S. Wood, *The Radicalism of the American Revolution* (New York: Knopf, 1992), p. 229. Wood's book is a continuing series of lessons in the displacement of one natural social artifact with another and the attendant problems, including that of those wedded to one natural social artifact being able even to comprehend those moving to the other. It is difficult for those used to thinking of themselves as social superiors to have everyone else treat them as equals. The difficulty is not just in being treated that way but in comprehending how anyone could treat them that way. See again Ronald Dworkin's discussion of courtesy as what I would call a natural social artifact (*Law's Empire* (Cambridge, Mass.: The Belpnop Press, 1986), pp. 47ff.).

7. Joel Feinberg, *The Moral Limits of the Criminal Law. Vol. I: Harm to Others* (New York: Oxford University Press, 1984), pp. 33ff.

8. This claim has the ring of analyticity to it, and that sort of ring is a warning to me to be cautious not to turn what I think is a useful analytical device, the concept of a natural social artifact, into an ideological and hardened metaphysical bludgeon.

9. Feinberg, *Limits of Criminal Law*, pp. 42–44.

10. It is not an irony, but only to be expected, that the demand for bottled

spring water, brought on in part by concerns about the purity of municipal drinking water, should lead to just such conflicts when those with access to springs sell the water to bottling concerns. In Callicoon, New York, Andrew J. Krieger is selling water from the spring on his farm to the Great Bear Spring Company, and the objection is that he is "selling one of the area's most precious resources, the crystal-clear water that bubbles up from an aquifer beneath his farm" (Lindsey Gruson, "Water Farm Stirs Tempest in a Bottle," *New York Times* [August 15, 1993], p. A39).

There are two objections. The first is that this has led to a loss of water in some of the area's streams, an objection that will be difficult to sustain given the recent drought in the area (ibid., p. A40). The second is that he is selling a common resource, the water from the aquifer that flows beneath the entire region, and that he has no right to sell it for his own profit, without regard to the rights of others to ensure that the aquifer be maintained, merely because he happens to own property where the aquifer breaches the ground in the form of a spring. This second objection is obviously more difficult to deny, in part because the worry about the aquifer being maintained is not an idle one considering that the region thrives in part because of the trout fishing in the streams the aquifer makes possible.

11. New York City's recent concern about the quality of its drinking water presents a nice illustration of how different interests can compete. The increasing population in the Catskill Mountains has caused a change in the quality of the runoff into the system that serves New York City's reservoirs. The City is concerned about salt being used on the increasing number of highways, as we have seen, about the human waste being produced by more and more residents, about the increased use of fertilizer in farming, and about the increased amount of farm waste being produced. So changes would impact on other interests people have besides those in the quality of their drinking water.

Such changes "could affect building in upstate New York by increasing the minimum distance septic systems may be built from streams in the reservoir watersheds. Also, public and private sewage treatment systems could have to build underground discharge systems, because discharge into the watershed could be banned. Agriculture could be altered by increases in how far animal and other farm wastes have to be stored from waterways, as well as increases in how close manure or fertilizer can be applied to land near sensitive waterways" (Allen R. Gold, "New York City Is Proposing Tighter Water-Quality Rules," *New York Times* [September 12, 1990], p. B3).

The list of possible effects could no doubt be lengthened and illustrates how difficult it is to do anything regarding waste, for instance, without setting back the interests of some who will have no clear gain: No one is claiming that those who live in the Catskills are going to have higher quality drinking water if the changes are made, even though such changes may make their general environment healthier.

12. Ground water is at serious risk of being contaminated by pollutants from abandoned oil wells. Among all the ways in which that pollution can occur, Ken Kramer, director of the Texas Sierra Club is quoted as saying, "'the seepage of

brine from unplugged wells into fresh water aquifers can be the most difficult to identify and clean up because it is hidden pollution, taking place underground.'

"In the oil patch of Oklahoma and West Texas, brine formations lay at depths of 1,000 to 8,000 feet, where they are under tremendous pressure. Many strata of virtually impermeable rock lay between the brine and the fresh water aquifers, which are near the surface down to depths of 500 feet." Cutting a well through the strata, and then abandoning the well, allows the brine to seep into the water aquifers (Robert Suro, "Abandoned Oil and Gas Wells Are Now Portals for Pollution," *New York Times* [May 3, 1992], p. A20).

13. A look at a map of the United States indicating where the water is tainted shows what does not take much thought to infer, namely, that the water is most tainted in those areas of the country where most people live and where more industries are concentrated. Ohio, Indiana, and Louisiana, for example, have more tainted water than Montana or Wyoming (John Holusha, "The Nation's Polluters—Who Emits What, and Where," *New York Times* [October 13, 1991], p. F10).

14. One understudied potential effect of environmental pollutants is "their role in disrupting hormonal systems of animals, particularly those governing reproduction" (John R. Luoma, "New Effect Of Pollutants: Hormone Mayhem," *New York Times* [March 24, 1992], p. C1).

"The effects of the chlorinated organic chemicals are most significant in fetuses [of the experimental animals]. They include partial retention of sex glands of the opposite gender, profound changes in sexual behavior and reduced fertility." The "use of the banned drug diethylstibestrol, or DES, may have served as an experiment" on human reproduction. Children of those women who were prescribed the drug "suffered from such effects as malformed reproductive tracts, infertility and a rare cancer called vaginal adenocarcinoma" (ibid., p. C9). So concerns are not just about cancer but also about other harmful effects on our health.

15. Exposure to dioxin increases the "risk of developing two types of lethal cancers, lung cancer and particularly soft tissue sarcoma, a cancer of connective tissue." It has been found that the risk is clearly increased for those working with or exposed to high levels of dioxin, but "the levels of dioxin ordinarily found in the environment had not been shown to be dangerous to people" (Keith Schneider, "Panel of Scientists Finds Dioxin Does Not Pose Widespread Cancer Threat," *New York Times* [September 26, 1992], p. 9).

The heading of this article is misleading. There is a difference between finding something not dangerous and not finding something dangerous. The panel of scientists did not find dioxin dangerous to humans even though they found "that exceedingly minute levels of dioxin caused biological havoc among fish, birds and other wild animals" (ibid., p. 9). Not finding dioxin dangerous does not mean that it is not. On the one hand, it may be that it will be found to be dangerous, that all the evidence is not in. And, indeed, the panel's final findings are not in. On the other hand, the *assessment* of danger is itself at issue. No finding of actual harm to humans by exposure to "dioxin ordinarily found in the environment" does not support an inference of no risk of real harm, it could be argued, if biologically complex animals can be harmed significantly through exceedingly small doses.

What is at issue is how to assess the risk: If complex animals can be harmed, then it is not unreasonable, it could be argued, to *presume* that humans are at risk of being harmed. To deny the presumption is to make an alternative presumption of how to assess risk, and that presumption needs to be argued for.

16. It is startling that homeowners "use **10 times more chemical pesticides per acre than farmers do**" (Malcolm Jones, Jr., "The New Turf Wars," *Newsweek* [June 21, 1993], p. 63; bold print in the original). See also Herbert Bormann, Diane Balmori, and Gordon T. Geballe, *Redesigning the American Lawn: A Search for Environmental Harmony* (New Haven: Yale University Press, 1973).

17. The amount of lead that can leach from dishes to a user varies considerably depending on "the heat and acidity of the food in contact with it and the duration of that contact. Tomato sauce will cause more lead to leach than mashed potatoes . . . and hot coffee will cause more lead to leach than water" ("Ceramic Dish Makers Agree to Curb Use of Lead," *New York Times* [January 17, 1993], p. A25).

18. In 1992, the Environmental Protection Agency (EPA) lifted a ban on certain farm chemicals. Ethylene bisdithiocarbamate (EBDC) can be used on "apples, barley, broccoli, cabbage, lettuce, cucumbers, rye, squash, watermelon, dry bean, eggplant and 22 other crops." The chemicals that use EBDC are "among the most versatile, least expensive, and most effective fungicides" (Keith Schneider, "E.P.A., in a Reversal, Lifts a Ban on Farm Chemicals," *New York Times* [February 14, 1992], p. A14).

The EPA argued that new evidence allowed it to lower its "estimate of EBDC's potency in causing cancer" when on that produce. Little or no residue was shown in the new tests. But EBDC breaks down into ETU, or ethylene thiourea, which is "among the most potent carcinogens used in agriculture" (ibid., p. A14). So one might argue that because the harm that one risks is so great, then even though the risk may be low, EBDC should continue to be banned despite the new evidence. This is what I would argue. We should have particular cause for concern given how often what was thought clear evidence later turns out to have been unclear.

19. "Atomic Waste Reported Leaking in Ocean Sanctuary Off California," *New York Times* (June 7, 1990), p. A12. Our present information regarding these barrels is not reliable. They are *reported* to have in them plutonium, mercury, and cesium, but a $900,000 study is being run to determine for sure what is in them. Edward Ueber, the manager of the Marine Sanctuary, said that "the original thinking was that it was a safe out-of-the-way environment to store these wastes," but the Marine Sanctuary supports the largest population of seabirds south of Alaska and "is one of the most productive fisheries on the West Coast, an abundant source of snapper, sablefish and herring" (Katherine Bishop, "U.S. to Determine if Radioactive Waste in Pacific Presents Danger," *New York Times* [January 20, 1991], p. A21). Either no one predicted when the decision was made to dump the barrels that they would corrode and leak, or if that prediction was made, it was not thought that any leakage would cause a problem. As I later argue, either possibility is difficult to sustain. So it may be that the problem with dumping such barrels in the sea was not that *firm* predictions could not be made but that no one thought any predictions were relevant. In any event, we have with this case

an object lesson for our inability to take firm predictions into account, and it is easy enough to suggest examples of our inability to make firm predictions, as with chlorofluorocarbons (CFCs), as we will see.

20. That it is reasonable is one reason why it is newsworthy that the space shuttle engines have been found to be more likely to explode than previously thought. The claim is that "the main engines [are] so temperamental that the risk of a catastrophic failure during a shuttle's fiery ascent to orbit [is] 1 in 120; earlier NASA estimates put it at 1 in 171" (William J. Broad, "Report Says Shuttle Engines Are Less Safe than Believed," New York Times [August 14, 1993], p. 6). So it is reasonable to make "improvements that will greatly enhance engine reliability and safety" (ibid.).

Perhaps more germane to our concern with environmental issues is that Jim Baca, "the director of the Federal agency that regulates the Trans Alaska Pipeline," is taking steps to ensure that an accident does not happen because, as "Richard L. Olver, the chief executive of British Petroleum Exploration," one of the three oil companies with an interest in the pipeline, put it, "Current performance falls short of the standards we all expect, and we are disappointed in this performance." Mr. Baca was blunter: "The reason that a disaster has not happened is not because of what the companies have done, but because of what individual employees are doing to keep this thing glued together." The problem is not just that "the risks can be serious or even catastrophic. In many cases, so little is in place in terms of systems, programs and accompanying data that the risks are indeterminate—no one really knows how severe the effects of such lack of controls could be" (Agis Salpukas, "Alaska Pipeline Risks Breaks, House Panel Is Told," New York Times [November 11, 1993], p. A14). But what we all do know is that faced with indeterminate risks that could be catastrophic, we ought to act to minimize the possibility of their occurrence. That is what Mr. Baca is doing—quite reasonably.

21. See "Cheap Chemicals and Dumb Luck," Audubon, Vol. 78, N. 1.

22. I do not deny, of course, that the state may have delayed timely action just so action would come to be inappropriate. This issue is, unfortunately, always in play about bureaucratic responses to problems that, without a timely response, will dissipate.

23. For a brief and excellent survey of the problem, see Paul Brodeur, "Annals of Chemistry: In the Face of Doubt," The New Yorker (June 6, 1986), pp. 70–86. I share what I infer to be Brodeur's judgments about how we ought to act in the face of risk. The details about ozone that I cite can be found in this article, and Brodeur's general attitude about how we make assessments of risk and act on these assessments is one I share and try to articulate.

24. The chief manufacturer of CFCs is, or was, E. I. duPont de Nemours & Company, and it maintained that "there was no reliable evidence that chlorofluorocarbons posed a hazard to ozone—or, for that matter, that the chain reaction worked out by Rowland and Molina [to explain how ozone depletion occurs] could occur at all. A duPont official testifying before the Subcommittee on Public Health and Environment declared that until there was actual proof to support the ozone-

depletion theory government regulation of chlorofluorocarbons was unwarranted" (Paul Brodeur, "In the Face of Doubt," p. 73).

DuPont is now phasing CFCs out of production at an even faster rate than it had originally planned to. The target date is the end of 1994 rather than the end of 1995 ("DuPont Hastens Chlorofluorocarbon Phase-Out," *New York Times* [March 9, 1993], p. D3). As we have seen, however, the EPA has asked duPont to delay phasing out its production of CFC-12 until the end of 1995 (Julie Edelson Halpert, "Scarcity of Car Coolant Could Prove Costly," *New York Times* [December 26, 1973], p. F5).

25. Similar but smaller decreases have been found over the Arctic. In January and February 1989, levels of ozone "near the North Pole were 25% lower, on the average, than in surrounding regions." In the Antarctic ozone losses have been "as high as 90% at high altitudes," and "in the Arctic the highest losses measured were 35%, the lowest 12%, with 25% the average." The decreases are related to vortices that form over the poles, and the vortex over the Arctic does not last as long as the vortex over the Antarctic (William K. Stevens, "Ozone Losses in Arctic Are Larger than Expected," *New York Times* [September 6, 1990], p. A25).

The most recent data indicate that things are even worse than they have been. The area depleted this past year was "9 million square miles." Antarctica "has a surface area of 5.4 million square miles," and this year's depletion did not match in area last year's, which was 9.4 million square miles. But the *magnitude* of this year's depletion was the worst ever recorded. "Ozone is measured in Dibson units, each of which represents the physical thickness of the ozone layer if it were compressed at the Earth's surface." The norm is 300 Dibson units, and the recent recording of "88 Dibson units . . . is the lowest value ever measured, anywhere in the world" (John Noble Wilford, "Antarctic Ozone Hits Record High," *New York Times* [October 19, 1993], p. A23).

26. "No one has the slightest way of knowing . . . what amount of ozone depletion is required to produce an important shift in the climate of the earth" (Brodeur, "In the Face of Doubt," p. 81).

27. Dixie Lee Ray, for instance, has argued that "until the transport mechanism for CFC molecules has been explained and until the full spectrum of CFC breakdown products is identified and measured *in the stratosphere*, the case for CFC destruction of ozone has not been made out" (Ray, with Lou Guzzo, *Environmental Overkill: Whatever Happened to Common Sense?* [Washington, DC: Regnery Gateway, 1993], p. 35).

I would argue that we do not need to know the mechanism in order to act, given the magnitude of potential harm and the current evidence for CFCs being the culprit. But Wieger Franson points out that although there has been skepticism that "CFC's emitted mainly by Europeans and Americans" could cause an ozone hole over the Antarctic, the breakdown of CFCs occurs only "in higher regions of the atmosphere," that transport to that elevation is relatively slow, that the winds encountered on that slow upward movement "transport a part of the CFC's emitted in the Northern Hemisphere to the Southern Hemisphere," that the temperature is so cold there that "there is little influx of ozone-rich air from near the equator," that the CFCs break down "when the sun comes up at the South Pole at the beginning of the Antarctic spring" and that the chlorine and bromine "freed from the CFC's" react with ozone "on the surface of polar stratospheric ice clouds, which only occur if temperatures are lower than −60 degrees" (Letter to the Editor, *New York Times* [October 12, 1993], p. A22).

28. We now know that the loss is not limited to the ends of the earth. It also occurs "throughout the Northern Hemisphere temperate zone, which includes North America, Europe, [and what was] the Soviet Union and most of Asia." The losses are "proceeding twice as fast as scientists had predicted," with "declines measured in the late fall, winter and early spring . . . [of] 4.5 to 5 percent in the last decade." Earlier estimates were that the ozone layer would be depleted by about 2 percent. William K. Reilly, then the administrator of the EPA, called the information of the more rapid depletion "stunning." "It is unexpected," he said, "it is disturbing, and it possesses implications we have not yet had time to fully explore" (William K. Stevens, "Ozone Loss over U.S. Is Found to Be Twice as Bad as Predicted," *New York Times* [April 5, 1991], p. A1 and D18).

Areas in Northern Europe, including Sweden, experienced losses of 8 percent, and projections of future losses of 10 to 12 percent over the next 20 years have already been overtaken as far too conservative ("A Bigger Hole in the Ozone," *Newsweek* [April 15, 1991], p. 64).

The Southern Hemisphere faces similar problems. The hole over the Antarctic stretches north into "areas like Punta Arena, Chile, and Ushusiah, Argentina," spiking hundreds of miles north (Nathaniel C. Nash, "Ozone Depletion Threatening Latin Sun Worshipers," *New York Times* [March 27, 1992], p.A7; see also Nathaniel C. Nash, "Unease Grows under the Ozone Hole," *New York Times* [July 23, 1991], p. B7).

29. Malcolm W. Browne, "Saving Ozone May Prove to Be Costly," *New York Times* (July 1, 1990), p. A9.

30. Ibid.

31. Chlorofluorocarbon 12 (CFC-12) is used as a refrigerant, and chlorofluorocarbon 11 is "the agent that puffs up the plastic foam that insulates the refrigerator's walls and provides much of the structural support of the box, because it is a good insulator" (John Holusha, "The Refrigerator of the Future, for Better and Worse," *New York Times* [August 30, 1992], p. F3). Replacing CFCs means finding both a new agent for the insulation and a new refrigerant, or some mechanism for cooling that is completely different.

32. Pollution knows no national boundaries. The Arctic now has haze, and "at times during the winter and spring it rivals pollution levels in large cities." It has been traced to industrial pollutants in the Northern Hemisphere, especially in Asia and Europe ("Tracing Arctic Haze," *New York Times* [January 15, 1991], p. C6). This haze now seems to be abating, diminishing by about "half during March and April, its worst months, since a peak in the early 1980's." The hypothesis is that this decrease is "linked to sharp reductions in pollution emissions in Europe and

the former Soviet Union" (Jon R. Luoma, "Sharp Decline Found in Arctic Air Pollution," *New York Times* [June 1, 1993], p. C4). The news is good, if true, but it is as much an indication of how sensitive the Arctic is to pollution produced elsewhere as it is of how the problem may be waning somewhat.

Another telling example of how extensively pollution can travel is found in the snow in Greenland, which is heavily laced with lead deposits that have now been shown to come from "leaded gas emissions from the United States" ("Lead in Greenland Is Traced to the U.S.," *New York Times* [August 17, 1993], p. C6). The lead levels have decreased significantly since "the advent of catalytic converters and unleaded gasoline," but, again, this is as much a mark of the sensitivity of the world to local pollution as it is a source of relief that the problem is diminishing somewhat (ibid.).

33. It seems as though almost everything about CFCs is a source of surprise. "The buildup of the industrial chemicals most responsible for deleting the earth's protective ozone layer has slowed substantially," it is now reported, and as "Dr. James W. Elkins of the National Oceanic and Atmospheric Administration's climate diagnostics laboratory" put it, "We were all caught off guard" (William K. Stevens, "Scientists Startled by a Drop in Ozone-Killing Chemicals," *New York Times* [August 26, 1993], p. A1).

The main culprits are CFC-11 and CFC-12, and it was found that "the growth rate for concentrations of CFC-11 dropped from an annual average of 11 parts per trillion from 1985 to 1988 to 3 parts per trillion in 1993. For CFC-12, the growth rate dropped from 20 parts per trillion in 1985–88 to 11 parts per trillion in 1993" (ibid., p. A18). Of course, this is a drop in the rate of *growth*, not a drop in the total amount, but the surprising news is that there has been any drop at all, and the hope is that these statistics are not a fluke and that "the worst of the ozone destruction [will now] come around the turn of the century" and things turn around from then on (ibid., p. A1). The turnaround will be slow, of course, because, it is claimed, it will take 50 to 100 years for things to right themselves again ("A Ray of Hope For the Ozone," *New York Times* [August 27, 1993], p. E2).

34. The BBC reported in early 1988 that the Soviet Union was asking for more scientific research on the problem. They presumed the need, I assume, of more information even in the face of great risk.

35. Brodeur, "In the Face of Doubt," p. 77.

36. The "natural" reaction to the risk of acting without all relevant information "might be to 'wait-and-see,' deferring any corrective action until it is clear that something detrimental is happening," argues Claud S. Rupert about ozone depletion. "But," as he adds, "this course has its own risks. The slow transport of materials from the troposphere into the stratosphere means that waiting for signs of visible trouble will load up the troposphere with a larger quantity of material to leak into the stratosphere afterward, thereby increasing the difficulty should it then be desired to stop the pollution. In the case of long-lived materials such as CFCs one would be committed to a final ozone depletion somewhat larger than that existing when corrective action was initiated, followed by a very leisurely re-

covery" (*The Analysis of Actual Versus Perceived Risks*, [New York: Plenum, 1983], p. 319).

37. The changeover in regard to aerosol spray cans began in the 1970s, and it has been claimed that "by the time the Environmental Protection Agency and Food and Drug Administration banned chloroflurocarbons from spray cans in 1975, the great majority of products were using alternative propellants" (Carol Lippincott, Letter to the Editor, *New York Times* [September 14, 1993], p. A24).

38. "CFC's have [now] been largely eliminated as spray propellants, as cleaning solvents and as ingredients in plastic-foam products like drinking cups and packing materials. That accounts for roughly half the original market, said F. Anthony Vogelsberg, the environmental manager of duPont's fluorochemicals division" (Stevens, "Scientists Startled by a Drop In Ozone-Killing Chemicals," p. A18). These changes could have been made quite some time ago without any undue burden on the industry, and we would arguably have that much less ozone destruction than we now face.

39. Browne, "Saving Ozone May Prove to Be Costly." p. A9.

40. An additional cost to manufacturers is associated with the risk of using a chemical that itself may have harmful effects, especially on the refrigeration units themselves. Simply replacing one substance with another as the coolant may have implications for the efficiency and life of refrigerators. It is "unclear whether lubricants can be developed that will be compatible with [a] new refrigerant and keep internal parts working smoothly for decades. And engineers at Whirlpool worry that [a] new fluid will gradually corrode tubing and components, possibly prompting massive recalls." In short, refrigerator manufacturers must design a new refrigerator without the extensive testing of its components that will guarantee reliability, and the risk is that a massive recall will be necessary in a few years that could bankrupt a manufacturer (Holusha, "The Refrigerator of the Future, for Better and Worse," p. F3).

41. Of course, even recent actions give no guarantee that we have turned the tide, for "it is pure speculation how quickly the ozone layer will recuperate without massive remediation efforts after the emissions have largely ceased" (Jeffrey Holman, Letter to the Editor, *New York Times* [September 14, 1993], p. A24).

42. Had we known we were short of the mark in 1975, of course, we would have had even more reason to act then. It is one of the implications of our epistemic state regarding such matters and of our track record of underpredicting that when the magnitude of harm is great we ought to presume that we are far *less* knowledgeable than any such predictions as those we have made regarding the ozone have made us appear. I say this with full regard to its irony.

As one would expect regarding such matters, we are still short of the mark in our predictions. As we saw, scientists have been surprised by a drop in the growth of CFC-11 and CFC-12. But they have also been surprised to find that ozone levels over the Northern Hemisphere are at their lowest in fourteen years, "down 10 to 20 percent from their normal range in the middle latitudes of the Northern Hemisphere" ("Northern Hemisphere Ozone at 14-Year Low," *New York*

Times [April 23, 1993], p. A26). In addition, the ozone-destroying chemicals "remained in the atmosphere over the Northern Hemisphere much longer this winter than last" (Warren E. Leary, "Satellite Finds Growing Threat to Ozone," *New York Times* [April 15, 1993], p. B7). There is disagreement about why this is occurring.

43. One advantage they have is that "they break down more rapidly than CFC's in the lower atmosphere" (Malcolm W. Browne, "Doubts Ease on Protecting Ozone," *New York Times* [January 11, 1994], p. C5).

44. John Holusha, "Ozone Issue: Economics of a Ban," *New York Times* (January 11, 1990), p. 28.

45. In the search for new ways to clean circuit boards without using CFCs, electronics companies have developed two that save money. One uses a "mild solution of adipic acid, an industrial chemical, to remove oxides from metal surfaces before soldering," and the soldering takes place in an inert atmosphere. The other involves switching to a flux that is 98 percent liquid so that there is little residue left on the boards, and most of what is left boils off as a circuit board is heated after soldering. "The little residue that remains will not affect the projected 40-year life of the equipment" (John Holusha, "Making Circuits without CFC's," *New York Times* [December 18, 1991], p. D7).

Both of these methods save the companies money. For instance, the no-clean method means that the company no longer has to buy the cleaning chemicals, handle them, or reclaim them and dispose of them. It should be noted as well how quickly these alternatives were developed.

In addition, we have to consider that manufacturers have adopted certain processes that now cause problems when other processes were available. Refrigerators in Japan, for instance, have long been absorption chillers, which use much less electricity and require no CFCs as coolants (John Holusha, "Chilling without Ozone Damage," *New York Times* [March 4, 1992], p. D7). Of course, manufacturers there have been encouraged to produce such refrigerators by a policy that discourages use of electricity whereas American manufacturers did not, and do not, have to face such a policy. We have here a good example of how such national policies regarding energy consumption affect manufacturing choices. That manufacturers chose a method they need not have chosen should be taken into account when assessing whether compensation is necessary and how much, but we also must take into account that they were not compelled to choose any other method because of our national policies regarding energy consumption. We have with this case a useful way to extend the analogy of going into a black hole to weigh one scientific claim or prediction against another. Determining liability for current policies for purposes of compensation is also to enter a black hole of claim and counterclaim. It is not obvious that any objective answer would ever emerge.

46. Finding alternatives to CFCs in refrigeration has been more difficult than finding them for cleaning circuit boards, but the work is under way. Utilities have had a contest for a design that would use 25 percent less electricity than the government standard and uses no CFCs. They received over 500 responses (Matthew L. Wald, "Utilities Go to the Source for Efficient Refrigerator," *New York Times* [Octo-

ber 17, 1992], p. 25). Whirlpool and Frigidaire were the finalists (Matthew L. Wald, "Whirlpool and Frigidaire in Refrigerator Contest," *New York Times* [December 8, 1992], p. D5).

One of the more likely alternatives is what is called a thermoacoustic refrigerator. Its underlying principle is that a loudspeaker sets up a standing wave, "in which individual molecules of gas are in rapid oscillating motion while the wave as a whole does not move at all." As the molecules move, they can carry "heat away from the thinner regions of the standing wave toward its thickest plate. If a metal plate is placed within this antinode, the plate absorbs heat from the gas molecules and can conduct it away from the refrigerator, leaving cooled gas in its place"—the refrigerator as pipe organ, so to speak (Malcolm W. Browne, "Cooling with Sound: An Effort to Save Ozone Shield," *New York Times* [February 25, 1992], pp. C7). The difficulty with this alternative is the difficulty with any innovative design, namely, cost. It would require removing expensive equipment already in place and "investing in entirely new systems" (ibid., p. C1).

One new suggestion is that a refrigerand like water be used to absorb heat and that heat then be "absorbed by another fluid, like lithium bromide" that "quickly [dissipates] the accumulated heat. A modest pump is enough to keep the process going" (Holusha, "Chilling without Ozone Damage," p. D7). The proposal gets rid of the need of CFCs as the liquid of choice and the need for an energy-hungry compressor. This solution solves another problem caused by replacing CFCs. Compressors require lubricating oils, and although CFCs are compatible with oil lubricants, it is not obvious that replacing CFCs with some other liquid and maintaining the current compressor design will produce a reliable machine. One solution to this problem would be a compressor that does not need lubricating oils, and a prototype of one, produced by Sunpower, has been developed that would also, it is claimed, reduce the need for electrical energy by 9 to 15 percent (John Holusha, "Rethinking the Refrigerator," *New York Times* [June 27, 1993], p. F10).

These alternatives leave untouched the problem of CFCs used in the insulation, but that problem seems easier to overcome ("Can Glass Keep the Fridge Cool?," *New York Times* [April 21, 1991], p. F7).

47. *International Herald-Tribune* (April 3, 1989), p. 3. One good outcome of the oil spill in Prince Williams Sound is that some protection will eventually be in place for furture spills. The Oil Pollution Act of 1990 requires that "shippers be able to provide 10,000 barrels a day of skimming capacity within 12 hours of a spill; 20,000 barrels of capacity within 36 hours, and 40,000 barrels within 60 hours." Unfortunately, a major tanker "can spill millions of gallons in a day or two" (Matthew L. Wald, "New Oil-Skimming Ship Prepares for Next Spill," *New York Times* [March 27, 1993], p. 7).

The oil spill off the beaches of San Juan, Puerto Rico, was significantly less damaging than it could have been because of "the presence on the island of much of the cleanup equipment needed to attack and contain the spill quickly"—a result of the legislation passed after the *Exxon Valdez* spill requiring "the stockpiling of cleanup and pollution-control equipment at 19 sites, including San Juan" (Ronald Smothers, "Stockpiled Equipment Aids Cleanup of Oil in San Juan," *New York Times* [January 9, 1994], p. A12).

48. In a recent meeting about setting international limits on carbon dioxide omissions, American officials were quoted as saying that their concern to negotiate a treaty that would aggressively reduce carbon dioxide emissions to combat global warming was not shared by the White House. "Mr. Sununu, said the officials, believes that without more conclusive scientific evidence, the United States should not commit itself to reduce emissions of carbon dioxide because such reductions are tantamount to reducing the use of oil and other fossil fuels and could lead to economic stagnation" (Keith Schneider, "U.S. to Negotiate Steps on Warming," *New York Times* [February 15, 1991], p. A7).

49. Recent evidence suggests that Prince William Sound was not all that pristine when the *Exxon-Valdez* went aground. In fact, "it is easier to find residue from an unnoticed 1964 spill than from" the *Exxon-Valdez*. That residue is "significantly richer in carbon 13 than the Valdez residue." The evidence suggests that it comes from the Gulf of Monterey in Southern California and that it "came from a spill related to a huge earthquake in 1964, which damaged storage tanks in the Port of Valdez and was the only event large enough to account for such a previously unknown spill that occurred nearly three decades ago" (Agis Salpukas, "A New Slant on Exxon Valdez Spill," *New York Times* [December 1, 1993], pp. D1 and 7).

Such a discovery will complicate determinations of liability, but ought to make no difference to the argument I am making here. That Prince William Sound was not *completely* pristine before an oil tanker terminal was sited there should not modify our reluctance to risk the additional harm that any major oil spill would cause. Indeed, that there is such residue from an earlier spill is further evidence of the harmful long-term effects of such disasters.

The oil does not disappear, but persists for long periods of time after it is spilled. In some situations, we ought to think of the problem as "a chronic oil spill," according to Dr. Jeremy B. C. Jackson. For the spill does not just occur and the oil settle, never to cause further harm. The oil from a spill off the coast of Panama "killed expanses of sea grasses, and with them their root systems, as well as some 150 acres of mangrove trees and their root systems." But after settling in the sediment in the water, the oil flushes up again each spring when the rain waters flood into the lagoons and kills whatever has taken hold since the previous spring rains. The persistence of the oil means that the harm will last for many, many years (Jon R. Luoma, "Some Oil Spills Repeat Harm Again and Again," *New York Times* [December 21, 1993], p. C4).

50. William K. Stevens, "Skeptics Are Challenging Dire 'Greenhouse' Views," *New York Times* (December 13, 1989), p. A18.

51. William K. Stevens, "Scientists Confront Renewed Backlash on Global Warming," *New York Times* (September 14, 1993), p. C1.

52. William K. Stevens, "Carbon Dioxide Rise May Alter Plant Life, Researchers Say," *New York Times* (September 18, 1990), p. C1.

53. Ibid., p. C9.

54. William K. Stevens, "Warming of Globe Could Build on Itself, Some Scientists Say," *New York Times* (February 19, 1991), p. C4.

55. New studies indicate that mercury in the atmosphere is increasing at a steady pace. From 1977 until 1990, "mercury concentration increased at an annual rate of 1.46 percent in the Northern Hemisphere and 1.17 percent in the Southern Hemisphere" (*New York Times* [February 18, 1992], p. C2).

56. "New Cause Of Concern on Global Warming," *New York Times* [February 12, 1991], p. C6.

57. S. Fred Singer, Letter to the Editor, *New York Times* [September 28, 1993], p. A24.

58. Ray, *Environmental Overkill*, p. 6. See also Ben Bolch and Harold Lyons, *Apocalypse Not: Science, Economics, and Environmentalism* (Washington, D.C.: Cato Institute, 1993), p. 10. This vision of what is necessary for public policy decisions also underlies the argument of Patrick J. Michaels, *Sound and Fury: The Science and Politics of Global Warming* (Washington, D.C.: Cato Institute, 1992), p. xi.

59. Ray, *Environmental Overkill*, p. 43. It is, of course, to be noted that the harm emphasized is economic harm to current interests.

60. The Clinton administration has adopted this "middle path" in its response to the greenhouse effect. It has committed itself to reducing "emissions of so-called greenhouse gases to 1990 levels by the year 2000," in part by helping "industries speed the adoption, over the long term, of alternative energy technologies now on the cusp of economic practicality" (William K. Stevens, "U.S. Prepares to Unveil Blueprint for Reducing Heat-Trapping Gases," *New York Times* [October 12, 1993], p. C4; see also "Clinton Addresses Warming of Globe," *New York Times* [October 20, 1993], p. A20). It also intends to encourage efficiencies that will ultimately pay for themselves, such as capturing methane from pig manure "to burn . . . for heat and electricity" (John H. Cushman, Jr., "Promoting Home-Grown Ideas for Cleaner Air," *New York Times* [October 24, 1993], p. E4).

61. The stance I am suggesting gets support from consideration of a variety of environmental issues. Recent concerns about the loss of fish is a case in point. The magnitude of potential harm is great. The extinction of some species would seriously harm our food supplies, and the loss of some species to fishing because the depletion of the supply has already caused widespread harm, both to those dependent on fishing for a living and to those who formerly had a relatively inexpensive source of protein. The loss of species would have other effects, obviously, than simply diminishing our food supplies, but I put those to one side here. It is not that those effects are not important but that the magnitude is heavy enough without weighing those concerns in the balance.

But if the magnitude of potential harm is great, information might be claimed to be relatively scanty. It is hard to count fish in any event: They move. But, in addition, "some data are faked by fishermen who are trying to evade conservation measures by underreporting their catch," it is expensive to do a thorough job, data collected by one nation are not shared with other nations, and so on (David E. Pitt, "Despite Gaps, Data Leave Little Doubt That Fish Are in Peril,"

New York Times [August 3, 1993], p. C4). But enough evidence is available to suggest caution: Cod have virtually disappeared from the Grand Banks, pollack have disappeared off Russia's Pacific coast, "there were significant drops in catches of Pacific perch, Atlantic redfish, yellow croaker, Atka mackerel, Atlantic mackerel and Atlantic herring" from 1970 to 1989, and during "the same period, aggregate hauls of four other important species—Atlantic cod, haddock, Cape hake and silver hake—plunged from 5 million metric tons in 1970 to 2.6 million metric tons in 1989" (ibid.).

It is arguable both that the evidence is so obvious that it cannot be ignored and that the evidence is scanty enough that we cannot be sure of our figures—which may be an artifact of something else other than depletion of fish, for all we can know. But with the magnitude of possible and *actual* harm so great, it is the cast of mind that matters here. Caution in catching fish seems the appropriate response, and Russia's "demand for a three-year moratorium on fishing" pollock off its coast seems unreasonable only in that the time period may be too short.

What seems the reasonable response to the apparent sudden decrease in fish is, I think, evidence for my claim that in the face of harms of great magnitude, caution is the reasonable response—even without certainty about the likelihood of the harm's occurring. Some evidence that the cast of mind I am urging does operate is that new rules have been promulgated in response to these uncertainties. The take for some species is to be reduced "by 50 percent over the next five to seven years," and "fishermen and dealers will . . . be required to buy electronic equipment to help report catches" ("U.S. Approves Plan To Reduce Fishing Off of New England," *New York Times* [January 4, 1994], p. A8).

I would argue that it is enough that our *interests* in catching and eating fish may be set back by the decrease in fish, but it has been claimed by Mary Harwood, "a biologist and fisheries policy expert," that if "you wish to fish, then the exercise of that right must be completely matched with a duty to know what impact you're having, to contribute to the group analysis process of that stock, and to assist in creating one if it doesn't yet exist" (David E. Pitt, "Despite Gaps, Data Leave Little Doubt That Fish Are in Peril").

62. As with almost any claim in this area, there is disagreement about how long carbon dioxide will reside in the atmosphere. Some claim it will be there "for centuries" (Stevens, "Scientists Confront Renewed Backlash on Global Warming," p. C6). Others claim it has a resident lifetime of "about 40 years" (Singer, Letter to the Editor).

63. The magnitude of harm from the spill is obviously a subject of much dispute and will continue to be at least as long as significant monetary issues remain unsettled and no doubt long after. On the one hand, Exxon argues that at the present time "all the animal populations have essentially recovered" and that the original damage to such animals as "sea birds, sea otters and pink salmon was less than the Government says" (William K. Stevens, "New Data on Spill Offered by Exxon," *New York Times* [April 30, 1993], p. A18). But it has been argued that the harm was significantly greater than present judgments or judgments made at the settlement with Exxon might lead one to believe. There is "extensive damage to sea

otters, killer whales, harbor seals, seabirds and fish, a result not only of the spill's immediate effects but also of oil remaining in Prince William Sound," and the concern is that "some of the worst damage from the spill [will] not show up until years after the accident, as the effect of oil [works] its way through fish-spawning and animal-breeding cycles" ("Valdez Spill Toll Is Now Called Far Worse," *New York Times* [April 18, 1992], p. 6).

It is a charitable, but also true, reading of this dispute by Dr. Robert B. Spies, "the chief Government scientist," who says "that the sharply divergent assessments of damage to wildlife populations were partly a product of scientific uncertainty that allows for a variety of interpretations" (Stevens, "New Data on Spill Offered by Exxon"). Part of what is at issue is that Exxon wants to deny any connection between "this year's low return of pink salmon and the spill" because it wants to deny legal liability and moral responsibility for any further damage that is discovered ("Alaska Fishermen Blockade Tankers," *New York Times* [August 23, 1993], p. A8). So a cynic might be pardoned for thinking that Exxon's claims would be driven by legal considerations if not by "divergent assessments" because of "scientific uncertainty."

But whatever the causes of the dispute, it would be a mistake to premise any decision about what we ought to do on settling it. To enter into it is to enter a black hole of calculating probabilities and assessing risks with indeterminate evidence. I do not mean in making this point to suggest that one side is right and the other wrong but that we make a mistake if we allow ourselves to be sucked into this debate about what harm has really occurred and what has not *if* we are concerned to set public policy about such matters as whether to continue efforts to cleanse the Sound or to preserve any wilderness in the area from logging so as to ease pollution in the Sound ("Oil-Spill Trustees Back a Land Deal," *New York Times* [June 16, 1993], p. A21). Making a timely decision about such preservation is crucial, and letting it rest on settling an issue that will not be settled any time soon, if at all, is a mistake no matter what the outcome. The likely outcome of letting things take their course until the issue is settled is that logging would occur, and any damage would be done, but even if nothing happened and the trees remained, we would have lucked into the decision that ought to be made on the reasonable ground that the magnitude of potential harm to the Sound is too great to justify logging.

64. One way to ensure that predictions of global warming are more accurate is to test the computer models against past changes in the climate. One study "uses climatic data from two periods in the past, one 20,000 years ago, in the depths of the last ice age, and the other in the mid-Cretaceous period 100 million years ago, when the temperature was 18 degrees warmer than now." These data, it is argued, provide a test for the claim that "a doubling of the atmospheric content of [carbon dioxide] would raise the average global temperature by 4.5 degrees." The prediction is that "the global climate would warm by 5.4 to 7.2 degrees by the year 2100" (William K. Stevens, "Estimates of Warming Gain More Precision and Warn of Disaster," *New York Times* [December 15, 1992], pp. C1 and C9). Of course, the study has been subjected to criticism, and the main criticism is that usually

raised about computer simulations, namely, that the assumptions on which the prediction is based are themselves "very, very shaky."

One can draw samples of such predictions from almost anywhere in the literature over the past ten years or so, and this one is meant to be representative. There are many more, with different predictions, if one finds this one wanting (see Kim A. McDonald, "In Forest at Duke U., Researchers Prepare for a Glimpse of the 21st Century," *The Chronicle of Higher Education* [July 28, 1993], pp. A9 and A14). But disagreement is the most common feature one finds (see e.g. "Scientific Debate Continues to Rage over Rise of CO_2, Global Warming," *The Chronicle of Higher Education* [July 28, 1993], p. A14).

I do not, of course, mean to leave the impression that such studies are not helpful or should not be encouraged, only that we cannot claim, given the state of our ignorance, to have knowledge about whether or not the earth is getting warmer. One way of putting my concern is that we ought to be in a position to take reasonable action even when we are ignorant. We should not be condemned to inaction in the face of the possibility of real harm because the evidence is not yet all in about the likelihood of such harm.

65. Any variable is a source of problems for any complex system. It is easy enough to find examples of almost anything's breaking down, for instance, or of accidents occurring. At Fifth Avenue and Broadway in New York, a "backhoe accidentally struck [a] 20-inch steam main during repairs," "contaminating the immediate area with asbestos" (Ronald Sullivan, "Steam-Pipe Rupture Sprays Asbestos in Midtown," *New York Times* [June 26, 1993], p. 23). So just as we ought to presume that any natural social artifact will have untoward unintended effects, so we ought to assume that any complex mechanism, such as a landfill site, will be prone to some accident or other occurring at some time during its relatively long lifetime.

66. "In what turned out to be a financial disaster for the [fishing] industry in the 1980 season, a photograph of a sablefish swimming near the submerged barrels was published in newspapers around the world, and the Japanese market for that fish immediately collapsed. 'They canceled all orders for sablefish, not just ones from this area,' Mr. Ueber [the manager of the marine sanctuary] said. 'There was a $10 million decrease in sales in what had been one of the five major fisheries on the coast'" (Bishop, "U.S. to Determine If Radioactive Waste in Pacific Presents Danger," p. A21).

67. Matthew L. Wald, "Explosion Could Release Radiation at Hanford, U.S. Panel Warns," *New York Times* (July 31, 1990), p. A16.

68. Matthew L. Wald, "Nuclear Hazard Festers Years after Alarm," *New York Times* (December 24, 1992), p. A10; all further quotations in this paragraph are to this article and to p. A10.

69. Ibid.

70. Ibid.

71. We also have "millions of pounds of highly radioactive fuel . . . sitting in Energy Department storage pools for so long that they are rusting and spreading

radioactivity." The pools were designed to hold processed fuel "for 18 months or so, to cool until the plutonium and other valuable material could be extracted from it," but in the 1980s, safety problems with the processing plants caused the fuel to be "marooned" (Matthew L. Wald, "Uranium Rusting in Storage Pools Is Troubling U.S.," *New York Times* [December 8, 1993], p. A1). "About 94 percent of the fuel is in pools," some "more than 40 years old" that "do not meet earthquake standards" (ibid., p. A20). Some is "stored dry and safe from corrosion in . . . concrete-and-steel casks," and some is buried "in carbon-steel tubes that are probably rusting by now," and "the Energy Department is no longer sure precisely where" (ibid.).

Here we may wonder what those were thinking who buried the fuel and neglected to keep accurate records, but we ought also to wonder at what was going on in the minds of those who created the bulk of the problem through apparent neglect—a failure to do anything to mitigate the harmful effects of leaving highly radioactive fuel in storage pools that are bound to rust.

72. Keith Schneider, "Nuclear Complex Threatens Indians," *New York Times* (September 3, 1990), p. A9.

73. In a recent flurry about a possible leak of 7,500 gallons of radioactive waste, Dr. Harry Harmon, "the vice president for tank waste remediation at Westinghouse, said . . . that the leak was an 'undesirable event,' but he added: 'A tank leak into the ground is not an immediate health issue because the radioactivity has to make it to the ground water and migrate to the river before it could interact with the environment. Studies have shown it takes decades for that to happen'" (Matthew L. Wald, "Deadly Nuclear Waste Seems to Have Leaked in Washington State," *New York Times* [February 28, 1993], p. A31). It is not clear what basis Dr. Harmon has for the claim that it takes so long for such waste to enter the ground water, but even if it were true, and it seems unlikely, that it is *said* by the person in charge of correcting the leaks in the tanks does not raise one's level of confidence that the problem is being taken seriously.

I am suggesting that we ought to act on the presumption that given the harm that would occur were the waste to contaminate the ground water, as much as possible needs to be done as quickly as possible to prevent such contamination from occurring. We ought to act, that is, as though any leak *would* enter the ground water—especially when, as in this case, the issue of when it does is contentious.

It is such statements that make outside observers suspicious about whether those in charge of Hanford have the right stance toward the risk, and the suspicions are sometimes confirmed. "At the beginning of last year, a state inspector . . . found three separate leak-detection systems, all inoperable" at the tank with the "biggest potential for disaster" ("Better 'Urp' Than 'Boom!,'" *New York Times* [June 21, 1993], p. A16).

74. Ibid.

75. It is, of course, easy to say what should have been done regarding the Hanford Nuclear Reservation and significantly less easy to say what should be done now. It is a depressing sign of how complicated the problem is that despite the Energy Department's promising the State of Washington and the EPA that it would begin "solidifying the waste . . . on the last day of this century, Decem-

ber 31, 1999," it now wants, with "its management contractor, the Westinghouse Hanford Company," to eliminate all deadlines and start over again. One sort of problem they need to worry about is that they will uncover unexpected difficulties as they proceed. The Savannah River Site is an object lesson. After deciding to encase radioactive wastes at the Savannah plant and building a plant, "now five years late and $1 billion over budget," officials discovered that it "cannot run because when the wastes are sent to a chemical-processing plant for preparation, they give off ammonium nitrate." Because the waste at Savannah is thought to be cleaner, as it were, than that from Hanford, the presumption of surprises at Hanford is high (Matthew L. Wald, "At an Old Atomic-Waste Site, The Only Sure Thing Is Peril," *New York Times* [June 21, 1993], p. A16).

It might be some consolation that Westinghouse was rebuked "for its management of the Hanford nuclear reservation for the six months ending September 30, 1993." It is less of a consolation to discover that the "rebuke" consisted of its not receiving its usual bonus—$2.7 million for the last six-month period. It will still receive $5.4 (Matthew L. Wald, "Manager of No. 1 Nuclear Site Is Rebuked by U.S.," *New York Times* [December 9, 1993], p. A22).

76. Of course, in regard to the Hanford Nuclear Reservation, the likelihood of an accident is high, and if it encourages further action to *presume* the likelihood of an accident, we should make that presumption. Trying to discount such a likelihood is difficult when we hear of such things as that "a worker accidentally turned on a mixer pump inside a radioactive waste tank that is considered the greatest safety threat on the sprawling site" and that—and this is even harder to believe—"a Kaiser Engineers worker lowered a rock by rope [into a tank] to test a pipe for blockage." He then pulled it out hand over hand and "suffered radioactive contamination on his hands and clothing" ("Mishap Casts Pall over Big Atomic Waste Site," *New York Times* [August 15, 1993], p. A18).

77. Federal auditors quoted by Philip Shenon, "Despite Law, Study Finds, Water in U.S. Schools May Contain Lead," *New York Times* (November 1, 1990), p. A1.

78. Quoted by Barry Meier, "Ban Is Sought on Mercury in Latex Paint," *New York Times* (April 6, 1990), p. C1. Twelve out of thirty-five latex paint companies in Michigan use mercury, two in excess of the legal limits.

79. Ibid.

80. This is according to Richard Denison, a senior scientist at the Environmental Defense Fund (John Holusha, "Farewell to Those Old Printing Ink Blues, and a Few Reds and Yellows," *New York Times* [May 13, 1990], p. F9).

81. Or so says Fred Shapiro, head of Pro-Flex Consultants in Huntington, New York (ibid.).

82. Ibid.

83. One should never underestimate the resistance of consumers. A number of years ago the Campbell Soup Company introduced freeze-dried soup in small aluminum containers: The packaging could be recycled, the cans were tiny compared with the traditional soup can, and I thought the chicken noodle and onion soups at least were delicious. Production was stopped because, as a company

executive informed me, consumers assumed that they could not be getting very much soup in such little cans.

84. Anthony Ramirez, "Soap Sellers' New Credo: Less Powder, More Power," *New York Times* [February 1, 1991], p. D14. The quotation is a paraphrase by Mr. Ramirez of a statement by Jay H. Freedman, an analyst at Kidder, Peabody.

85. Peter Passell, "Economic Scene: Greenhouse Gamblers," *New York Times* (June 19, 1991), p. C2.

86. Ibid.

87. Although it is common, the insurance analogy may mislead because we buy insurance so that we can continue to live in flood plains or along a coast despite concerns about floods and hurricanes, not to diminish the potential harm. I owe this point to Hugo Bedeau.

88. See Al Gore, *Earth in the Balance*, (Boston: Houghton-Mifflin, 1992), pp. 295ff. for a discussion of some of the issues involved in a global response to the concern about the greenhouse effect.

89. See *Policy Implications of Greenhouse Warming* (Washington, D.C.: National Academy Press, 1991), pp. 47ff., for a listing of some of the ways in which we could act to mitigate the potential harm. It is instructive to run through the list and see how beneficial to some of our other interests acting to cut emissions could be. Homeowners should do this just after receiving a utility bill.

90. John Holusha, "Keeping a Gadget-Mad Nation Charged Up—and Safe," *New York Times* (December 22, 1991), p. F10.

91. Israel Shenker, "Novo-Nordisk Rides the Enzyme Tide," *New York Times* (January 27, 1991), p. F6.

92. Soybean ink costs about 5 to 10 percent more than petroleum-based inks and has some other difficulties, including a slower drying time that requires slower press runs, but it does reduce emissions, and "the amount of sludge that must be disposed of as hazardous waste" ("A New Ingredient for Many Papers: Soybean Ink," *New York Times* [March 23, 1992], p. D8).

93. Matthew L. Wald, "Windows That Know When to Let Light In," *New York Times* (August 16, 1992), p. F9.

94. "Motorists in the United States generate about 1.3 billion gallons of used oil a year. . . . More than half of the oil is burned for fuel, mainly in industrial furnaces. Another 32 percent is dumped in landfills, and only 2 percent is recycled." Converting that oil to high-quality gasoline and heating oil will diminish the need for imported oil, reduce ground-water pollution, and decrease the costs associated with getting rid of the oil (Thomas C. Hayes, "New Process to Recycle Old Motor Oil," *New York Times* [February 11, 1992], p. D4).

95. "Removal of Nitrates," *New York Times* (March 10, 1992), p. C2.

96. The technique depends on using "what looks like a dishwasher packed with plastic Wiffle Balls" coated "with a liquid solution that reacts to and splits off particular compounds from the fumes as the fumes are pumped through for cleansing" ("Coated Balls Remove Hazards from Fumes," *New York Times* [October 12, 1991], p. 43).

97. Carbon dioxide emission is not reduced but is "the amount the vegetable

that was used to make the oil absorbed while growing. So there is no net accumulation" ("The Buses Run On Canola Oil," *New York Times* [June 7, 1992], p. F9).

98. The list given is meant as a representative sample of the sorts of new processes and ideas about how to reduce waste and the harms associated with waste that have appeared within the last few years. It is also meant to indicate the scope these activities will have. Any process that reduces the billions of gallons of oil that make their way into the environment is a significant addition to the changes occurring in our production of harmful waste.

99. In "Washington County, in upstate New York, the incinerator is short about 300 tons of trash a day it needs, said James D. Willis, director of solid waste for the Warren-Washington Counties Industrial Development Agency. And in the town of Huntington on Long Island, an incinerator with a capacity of 900 tons a day has been taking in just 550 a day" (Sarah Lyall, "Suddenly, Towns Fight to Keep Their Garbage," *New York Times* [January 5, 1992], p. E14).

One consequence of the need for garbage is that some municipalities have required that waste produced there remain there. But North Hempstead's "flow control law" was declared unconstitutional, and if that is upheld, it is claimed, "it could deny the town some 95,000 tons of trash a year and even drive its Solid Waste Management Authority into bankruptcy" (ibid.).

100. The advantage of the two new procedures developed, as noted, is that they not only cut the amount of CFCs being released into the atmosphere but also save money for the manufacturers (Holusha, "Making Circuits without CFCs").

101. For a discussion of wrongs, see Feinberg, *Limits of Criminal Law*, pp. 34–35.

102. A more thorough answer to this question would require a more thorough analysis of the nature of rights, but I do not provide one because I do not think I need more than I have given here to make my thesis clear. A second reason for hesitating to provide a firmer footing by appealing to a theory of rights is that I think there is no consensus on their nature. Thus an attempt to provide a theoretical underpinning will itself provide grounds for doubt. The risk of not providing such an underpinning, of course, is that what I have provided will itself be subject to doubt until a theory justifying them is itself provided, but I do not think I have made any claims contentious enough and so dependent on some particular theory as to provide grounds for such doubt.

Another reason for not providing a more thorough analysis of rights is that I think the argument about how to proceed can be settled by appealing to interests alone, so that theoretical problems about rights would not affect the conclusions I want to draw.

103. To talk of weighing rights is to talk metaphorically, and the metaphor misleads. It implies we have some sort of scale that measures rights, but, in fact, I would argue, what is required to assess rights one against another is a theory that includes their places in a political scheme, and their "weight" becomes as much a matter of their place as their heft.

Adding a new metaphorical dimension—location—may only mix up matters as it mixes metaphors, but what needs to be captured in any extended analysis of assessing rights is that sets of rights operate in different spheres and that these spheres may clash as well. So we must distinguish civil rights from political rights, for instance, and when we try to weigh one within one sphere with another in another, we must realize that more is at stake than simply the weight each has within its respective sphere. But replacing these metaphors with a viable theory would take us beyond what we need to handle the difficulties we face here.

104. It is often thought that what uniquely defines and so distinguishes professionals are the obligations they have to their clients. Lawyers are thus obligated to keep confidences, nurses to care for their patients, and so on, but no one would suggest that someone in business is obligated to care for a customer, or even for a customer's interests.

The claim that a professional is privileged to practice in a state is not uncontroversial because it requires defining professionals not wholly in terms of special obligations but also in terms of other sorts of relations they may have. The argument for there being other sorts of relation than those marked by special obligations has two parts.

First, a coroner is a professional who is privileged to determine the cause of death. He or she may be obligated to do that as well, depending on the state laws, but a lawyer has neither that obligation nor that privilege. Second, what a client tells a lawyer is privileged information, information that the lawyer and no one else is privileged to hear. I have no such privilege, and anyone who told me what they told a lawyer, or a psychiatrist, cannot count on my having any special obligation not to divulge what I was told. Lawyers are obligated not to divulge what they are told *because* they are privileged to such information.

So if this comparison of the status of corporations to that of professionals is to be fully persuasive, it needs to be backed with a new understanding of what it is that marks out a professional as a professional, and that requires a thorough analysis of the kinds of relations professionals bear as professionals to clients and perhaps others. Their positions are determined by more than simply special obligations.

See in this regard, Michael Bayles, *Professional Ethics* (Belmont, Calif.: Wadsworth, 1989). Bayles's book is a standard text in the area and is structured around the various obligations professionals have. No other kind of relation is mentioned, and one can scan the titles of the chapters to see how pervasive is the concept that all professional relations are reducible to obligations—"Obligations and Availability of Services," "Obligations between Professionals and Clients," and so on.

105. Senator Max Baucus has argued that "manufacturers should take 'responsibility for the life cycle of a product. . . . Anyone who sells a product should also be responsible for the product when it becomes waste.'" Legislation to require that could either "require manufacturers to include the cost of disposal in a product's price or . . . levy a tax on manufacturers to pay for recycling." One incentive for such legislation is that the costs of recycling are being borne by municipalities and is significantly higher than whatever can be gained by selling what has been recycled (John Holusha, "Who Foots the Bill for Recycling?," *New York Times* [April 25, 1993], p. F5). One of the points of making manufacturers responsible for the life cycle of a product is to encourage design that will maximize the ease of recycling. Such a bill would have to be carefully designed itself to ensure that

manufacturers are encouraged to redesign what they manufacture. One of the complaints about a similar German bill is that by putting the burden too heavily on the consumer, Germany has not provided enough of an incentive to manufacturers (Ferdinand Protzman, "A Nation's Recycling Law Puts Businesses on the Spot," *New York Times* [July 12, 1992], p. F5).

In speaking about the enormous problems created by quickly outdated computers and computer products, Mark Greenwood of the EPA's Office of Pollution Prevention has said that we "believe that companies can make choices in the design and manufacture of computers that reduce the environmental impact, and open up great new areas for recycling" (Steve Lohr, "Recycling Answer Sought for Computer Junk," *New York Times* [April 14, 1993], pp. A1 and D13). Making computer companies responsible for the life cycles of their products, however short, would hasten such choices.

Requiring that manufacturers of tires pay for the life cycle of their product by, for example, paying for those tires that are returned to them would go a long way toward solving the problem caused by the 279 million tires a year being thrown away in the United States (Barnaby J. Feder, "Shrinking the Old-Tire Mountain: Progress Slow," *New York Times* [May 9, 1990], p. C1). At least then any tires thrown away would become a commodity for those willing to collect them and turn them in—just as bottles and cans are collected by those who want the deposit money. New York City, to give just one example, would certainly gain some leverage on the four million tires it is estimated are "discarded—legally and illegally—in the city each year" (Michel Marriott, "Junked by Night, Dead Tires Haunt New York," *New York Times* [June 17, 1993], p. A1). At least the manufacturers would have most of them back. That might provide an incentive for them to change the way tires are made "so that the disposal problem will take care of itself" (John McPhee, "A Reporter at Large: Duty of Care," *The New Yorker* [June 28, 1993], p. 80). How to do that is not obvious, of course, or there would be no reason to landfill something that contains "more than two and a half gallons of recoverable petroleum" (ibid., p. 73). One possibility is to use the tires for something else, such as "rubber-based road pavement" (Feder, "Shrinking the Old-Tire Mountain: Progress Slow"). Another is somehow to extract the energy from them (Matthew L. Wald, "Turning a Slew of Old Tires into Energy," *New York Times* [December 27, 1992], p. F8). In any event, if tire manufacturers had to take their tires back once they were used, the very people most familiar with tires would have a huge incentive to explore possible uses for them after they had served their function as tires.

106. There is obviously a longer story to tell here, for it is easy enough to imagine those who do need to be concerned about the quality of air—movie directors unable to obtain the right look for a film because the quality of air is so degraded that the pictures lack the crispness and clarity needed, or manufacturers of high-quality optical equipment having their products damaged by minute particles in the air being deposited on the glass as it is formed, and so on. What is of interest is that those who need clean air produce it themselves: Movie directors take their sets indoors or go to a different city, manufacturers produce clean rooms, cleansing the air, and so on. Unfortunately, that some manufacturers go to some pains

to cleanse the air they need is no guarantee that they do not release it as dirty and polluted, if not more so, than it was. Cleansing air, like cleansing water, costs, and without a guarantee that other manufacturers are equally burdened by such costs, no manufacturer has any economic incentive to diminish any pollution incidental to manufacturing.

107. Schneider, "The Environmental Fix with a Legion of Doubters."

108. In laying down the conditions under which we need to concern ourselves about distributive justice, David Hume says that "even in the present necessitous condition of mankind . . . whenever any benefit is bestowed by nature in an unlimited abundance, we leave it always in common among the whole human race, and make no subdivisions of right and property. Water and air, though the most necessary of all objects, are not challenged as the property of individuals; nor can any man commit injustice by the most lavish use and enjoyment of these blessings" (*An Enquiry Concerning the Principles of Morals* [Oxford: Oxford University Press, 1989], p. 184). The implication that Hume explicitly draws is that when such "necessitous" resources are scarce, we can commit injustice by the lavish use of them.

109. It is this principle that lies behind the government's attempts to put into place a system for restoring the Everglades and preventing any future degradation. As Don Carson, executive vice president of Flo-Sun, a sugar company in Florida, put it, "We know that we have got to deliver water to the system that is as good as if we weren't there at all" (Larry Rohter, "U.S. and Florida Growers Reach Pact on Everglades," *New York Times* [July 14, 1993], p. A12).

Of course, as we might expect in such a contentious matter, involving millions upon millions of dollars, any factual claim that might make a difference to the outcome is a matter of dispute, and some "growers do not see why they should be obliged to discharge water that is purer than the water entering their fields," arguing, in short, that the water they get is already polluted and they should not bear all the costs of purifying it (Peter Passell, "Economic Scene: The Everglades Pact Could Be a Model for Environmental Accord," *New York Times* [July 15, 1993], p. D2).

110. Various corporations that have been "sued by the Environmental Protection Agency" to clean up toxic waste from dump sites have tried a variant of this argument. They have tried "to draw other waste producers—from towns and cities to small businesses and individuals—into multimillion-dollar cleanup cases." In a Connecticut case involving 23 municipalities, a judge ruled that they were not liable to share the costs because they "had not been shown to have dumped any of the hazardous substances specified by the E.P.A. under the Federal Superfund Law." But such decisions must be made case by case because it has been ruled that municipalities are not exempt from the Superfund law (George Judson, "Towns Escape Cleanup Costs at Waste Site," *New York Times* [December 24, 1993], pp. B1 and B5).

111. I should add "whichever alternative is better, all things considered." I want to emphasize that the choices are not cost-free so that we must compare the advantages and disadvantages of each option and choose accordingly, and I want

to emphasize that the choosing is itself a political matter that should not be left up to the manufacturers or businesses involved.

I am thinking here of such examples as fast food stores selling hamburgers and French fries in packages that get tossed and end up, in the best of cases, in landfills, but there are many other examples. How quickly things can be turned around is shown by McDonald's trash volume being "down 90 percent" in three years after it dropped its "plastic foam hamburger box . . . in favor of a much less bulky paper wrapper" in 1990. It also has done such "other things" as "substituting brown, unbleached takeout bags made of 100 percent recycled paper for bleached white bags made of virgin paper" (Holusha, "Environmentalists Try to Move the Markets").

112. This is the beginning of what must become a complicated argument. For instance, as stated, it assumes that *any* degradation of the environment is a loss that must have a cost to be counted in, but, for one example, some losses may be so minor that it costs more to determine how much of a loss has occurred than the loss itself costs.

Clearly, however, we have not taken into full account the total costs of production if we have ignored gross environmental losses—water polluted, air degraded, and so on. And if we were to take such costs into account and change prices accordingly, some things that are inexpensive would become much more expensive, such as water, to name an obvious example. In this regard, see Peter Passell, "Rebel Economists Add Ecological Cost to Price of Progress," *New York Times* (November 27, 1990), pp. B5–6, and Marlise Simons, "Europeans Begin to Calculate the Price of Pollution," *New York Times* (December 9, 1990), p. E3.

113. It might be thought that using rechargeable batteries would represent a big gain in diminishing the introduction of toxic materials into the environment. After all, one rechargeable battery can take the place of 300 alkaline batteries, and American consumers use 2.5 billion batteries a year, over 90 percent of which are not rechargeable. But in addition to not holding their charge very long, rechargeable batteries contain cadmium, a carcinogen (Eben Shapiro, "New Life for Rechargeable Battery," *New York Times* [February 15, 1991], pp. D1 and D4). One solution is to have the manufacturers recycle the cadmium, and two companies are accepting rechargeables for recycling, although they must send them to East Asia because there is no recycling plant in the United States for cadmium. It is not much of a step beyond what has already been done voluntarily by these companies to require that such batteries be recycled and to provide an incentive by adding a deposit fee high enough to entice returns when the battery's rechargeable life is complete.

114. Changing in such a fundamental way the costs of products will have a disproportional impact on the poor. We can already see that any full-scale attempt to recycle, for instance, will have different effects depending on differing economic levels. In Staten Island, for instance, which has the best percentage of recycling in the city of New York, 18 percent of the garbage is being recycled, whereas in northern Manhattan, which includes Harlem, only 5 percent is being recycled. In Newark, which has a recycling rate above 50 percent, there is "poor participa-

tion in low-income, high density areas," and because collection costs for curbside pickup are high, the suggestion is that buyback centers in low-income areas would be both more efficient and less expensive. A nonprofit collection center in the Bronx has sold about $5 million in recyclable products since 1982 while paying about $800,000 to those who have brought in trash (Allan R. Gold, "The Poor Mainly Recycle Poverty," *New York Times* [December 30, 1990], p. E6).

115. We can do this as prudent shoppers, ensuring that we have less waste to get rid of than we would otherwise have. "An estimated 180 million tons of household refuse and institutional waste were generated in 1988," and that "is expected to rise to 216 million tons in 2000."

"Packaging accounts for more than a third of all solid waste, a share that has grown steadily for three decades" (Lisa W. Foderado, "Trying to Hold Down the Level of Garbage to the Merely Immense," *New York Times* [November 30, 1990], p. B16). The argument is not that one ought to do one's share at least to diminish that growth, although that is an argument we could give, but that we ought to act in our own self-interest to diminish the trash we each produce and that one way to do that is to take in less that must be thrown away. Packets of juice boxes with their own straws "are terribly over-packaged and aren't easily recycled." We ought to *pre*-cycle, as the expression goes, being "conscious of buying habits and their effect on the amount of trash going to ever-shrinking landfill space" and on the amount we have to carry out each week (ibid.).

I have studiously refrained from pointing out how the practices of consumers can affect our current natural social artifact in part because the weight of significant change must go, I think, on manufacturing processes. It will not change much to preach to those whose habits of use are conditioned and justified by the natural social artifact they accept and cannot help but be subject to.

But if the natural social artifact we now have changes, those habits of use will of necessity change as well. As one variable in a complex causal network, either those habits of use will change as a result of changes in what is available to them because manufacturers have altered what they produce or consumers themselves will alter their habits and be a causal factor in producing that change. It is not an unimportant point that self-interest urges us to reduce what we must haul to the curb.

116. The argument here is not that the manufacturing jobs that will be lost, if any, do not have value but that they have no *more* value merely because they are jobs that people now hold than the jobs that will be created by cleaning up the Bay. If they did have more value merely because people *now* hold them, then the jobs that were displaced because such manufacturing polluted the Bay have a heavier weight too, and we would have to so design our policy towards the Bay as to encourage their reintroduction. It is no small matter that the oyster fleet in the Chesapeake at one time "employed a fifth of everyone involved in fishing in America" (Tom Horton, "The Last Skipjack," *New York Times Magazine* [June 13, 1993], p. 33).

But an argument comparing present jobs to be lost with past jobs already lost and future jobs to be obtained quickly becomes bogged down in attempts to calculate competing weights of jobs and likelihoods of various kinds of job creation. So I do not give such an argument. I am instead arguing merely that we cannot rest a policy about how to treat the Bay on any claim that jobs people now have have some special weight merely because they are jobs people now have.

117. There is no doubt that the losses that can be calculated have been high and sufficient to weigh heavily against any interests that now exist that may be harmed by an attempt to cleanse the Bay. But I resist trying to give a dollar amount to the loss because such an attempt both fails to reflect the epistemic problems I have tried to emphasize and might be misconstrued as an attempt to quantify the wide variety of losses, as though the losses were only economic. We would then have to provide an estimate of the cost of swimmers being unable to swim in the Potomac River (by, presumably, calculating the gas consumed by driving elsewhere, the costs of building and maintaining municipal pools to replace the River, and so on), an estimate of the costs of the loss of eagles and ospreys and peregrine falcons, an estimate of the cost of the loss of the pleasures of fishing for rockfish (for which there has been no season almost a decade), and so on.

The very request for a dollar amount, in short, presupposes that we can squeeze all our losses into a single currency or else squeezes out of consideration all the losses that cannot be calculated in dollars and cents and that are still, for all that, of great importance to those who have enjoyed the Chesapeake in its prime. As I have noted, it is one of my underlying themes, though not one I have emphasized or argued for directly, that cost–benefit analyses fail to come to grips with many of our most important public policy issues, especially in regard to the environment, because the very mode of analysis presupposes that all our interests can be adequately captured by appeal to a single value. It is one source of tension in natural social artifacts that they reflect a *variety* of values—as spoken French, in emphasizing elisions, marks its preference for mellifluous speech over speaking precisely what is written, as occurs in German. I have no theory to displace cost–benefit analysis, but the structure of analysis I have provided for making reasonable decisions in situations of doubt is meant to accommodate a different form of assessment.

118. It is a long road back. "While oysters appear to be making a comeback this year after several exceptionally barren years, the catch is drastically down from the past—only a tenth of what it was at the turn of the century" (B. Drummond Ayres, Jr., "Stirrings of Hope in Redeeming Chesapeake Bay," *New York Times* [December 3, 1990], p. A20). Indeed, in 1993 the catch was only 166,000 bushels, "less than 1 percent of the catch a hundred years ago" (B. Drummond Ayres, Jr., "Hard Times for Chesapeake's Oyster Harvest," *New York Times* [October 17, 1993], p. A1).

There are also long-range concerns about the crabs. The catch has not declined too significantly, but, as elsewhere regarding oysters and clams, for instance, "watermen are working harder now to get the same catch" (Tom Horton and William M. Eichbaum, *Turning the Tide* [Washington, D.C.: Island Press, 1991], p. 231). And crabs are just one indicator of the overall decline in the well-being of the Bay.

There are success stories elsewhere. Long Island Sound seems to be cleaner than it was primarily as a result of "improvements in local sewage treatment," and in 1992 a few beds of clams were opened for collecting (Constance L. Hays, "Saving the Sound: Hopeful Signs," *New York Times* [October 11, 1992], p. A46).

119. It is thus difficult to take as an objective judgment the claims by Formosa Plastics that the discharges from the $1.3 billion expansion of its plastics factory in San Antonio Bay will have no adverse effect on shrimping in the Bay. The company would clearly lose money if it were unable to complete what it has begun, and it is claimed that 1000 new permanent jobs would be lost as well as the jobs required for the construction. A Texas agency has given the company permission to "discharge up to 15 million gallons of chemical-laden waste water a day once the expansion [of the plant] is completed, an eightfold increase over the plant's current discharge limit." The water "may include sulfuric acid, chlorine, ethylene dichloride and vinyl chloride," all "known to cause cancer, blood clots and liver and kidney damage when ingested in high concentrations." It does not make one sanguine to know that Formosa Plastics has "already been assessed large fines by both Federal and state environmental regulators for violating its existing permits" (Sam Howe Verhovek, "Shrimpers Feel at Bay over Plant Expansion," *New York Times* [June 20, 1993], p. A16), and that sort of evidence ought to weigh more heavily than any concern about loss of jobs—given the potential danger to the shrimping industry from such discharges and the subsequent loss of jobs dependent on it.

120. Traditionally utility rates are set so that if a utility sells less energy than it is projected that it should sell, its profits drop and if it sells more, its profits rise. A utility company thus has a strong disincentive to reduce consumption of energy: The less energy used, the less profit it is likely to make.

The incentive should be the other way around: The less it sells because of conservation, the more it makes. And recently a number of states have reversed "the traditional relation between profits and energy"—New York, California, and Rhode Island. In New York, conservation goals are set for utilities. Orange & Rockland is to reduce "energy demand among its customers by 1 percent a year for each of the next three years." If it succeeds in reaching that goal, it can raise its rates. All consumers will thus pay more per kilowatt hour, itself an incentive for individual consumers to save, but those customers who reduce their energy consumption will be better off: Their increases will not be as great.

The traditional method of regulating utilities is "tailor made for destroying conservation incentives," according to Ralph Kavanagh, "director of energy programs for the National Resources Defense Council in San Francisco." The reversal of the old relation between profits and energy consumption is tailor made to encourage conservation (Elizabeth Kolbert, "Utility's Rates Are Tied to Energy Conservation," *New York Times* [September 1, 1990], p. A10).

121. If companies and individuals are to be encourged to conserve by an increase in utility rates, for example, they need to be provided with products and alternatives that effectively contribute to conservation.

The EPA has initiated a program to encourage electricity efficiency in lighting. The savings are immense, and not just in the total amount of money spent for lighting. "If all businesses adopted commonly available high-efficiency lighting techniques, the agency said that about 11 percent of all the electricity used in this country would be saved and costs would drop by $18.6 billion a year. It also said that the national ouput of sulfur dioxide, the prime cause of acid rain, would fall 7 percent, and output of carbon dioxide, which is believed to cause global climate change, would fall 5 percent" (Matthew L. Wald, "E.P.A. Urging Electricity Efficiency," *New York Times* [January 16, 1991], p. D6).

It should be noted that such efficiencies can be achieved using existing technology. Nothing new need be created except the will to change, and providing an economic incentive is one way of doing that, although that may raise a question of subsidies for those too poor to afford the new rates.

122. The town of High Bridge, New Jersey, has done this, replacing its flat collection fee of $280 per household per year with stickers to be placed on cans or bags or other disposable items like stuffed chairs. The system seems to be working in High Bridge ("Pay-by-Bag Disposal Cuts Volume of Trash," *International Herald-Tribune* [November 28, 1988], p. 3). But we will probably need incentives to discourage citizens from dumping their trash elsewhere as trash becomes more expensive to get rid of and as people realize that they are paying for every bit they produce. It is some measure of the problem now that the Salvation Army is removing its containers "in eight Mid-western states" because the containers have been stuffed with trash. The Salvation Army in that region "spent almost $1.5 million in trash removal last year, up from $180,000 the year before." Major Charles Nowell, "who oversees the Salvation Army operations in the South," said, "We're one of the biggest trash haulers that any city or county has in its community" ("Tired of Trash, Charity Removes Drop Boxes," *New York Times* [September 5, 1993], p. A41).

123. A former director of a sewage-treatment plant in West New York was indicted for dumping raw sewage and sludge into the Hudson. He also is accused of filing false reports and of doctoring water sample records. The claim is that he "stood to gain financially from making the plant appear efficient because the plant was using a sludge-treatment process that [he] marketed through his own consulting firm" ("Ex-Sewage Chief Cited as Polluter of Hudson," *New York Times* [February 15, 1991], p. B2). The dumping was discovered when a new director took over and "noticed unusually low sludge-production figures" (Allan R. Gold, "Sewage Company Head Resigns after Indictment," *New York Times* [February 17, 1991], p. A43).

The use of the criminal sanction for pollution, for instance, raises some serious questions about being even-handed when some major polluters have deep pockets to pay lawyers to protect their interests and others do not and about whether it is appropriate or efficient to brand as criminals those who pollute. For a brief discussion of some of these issues, see Allan R. Gold, "Increasingly, Prison Term Is the Price for Polluters," *New York Times* (February 15, 1991), p. B6. On the one hand, the criminal sanction should be used sparingly, only when the conditions are exactly perfect for its being effective, and it is not obvious that polluting is a perfect target (see Herbert J. Packer, *The Limits of the Criminal Sanction* [Stanford: Stanford University Press, 1968], esp. Part III). On the other hand, some companies continually cause environmental problems. The barge that ran aground and spilled oil on the beaches of San Juan, Puerto Rico, belongs to a family from New Jersey that has again and again been found guilty of safety violations—with "3 groundings, 9 oil spills, 2 collisions and 21 cases in which the Coast Guard had assessed fines for safety violations," all in 1990 alone (Ronald Smothers, "Leaking Oil Barge Has Ties to Polluter in Jersey," *New York Times* [January 10, 1994], p. A8). Such persistent disregard for the public good seems to invite the criminal sanction.

It is worth noting that one serious problem concerns the selective enforcement of the law, especially in an era of budget cuts. An owner of one recycling firm in New York estimates that it cost him a half million dollars to meet the conditions of the Department of Environmental Conservation. He had to provide concrete floors, for instance, so that waste would not wash into ground water, but now finds that in Suffolk County only he and one other operator have State permits, and there are over 60 companies hauling waste. The operators without permits can charge significantly less, and he is being driven out of business. "When he complained to the state, he was told there wasn't staff to enforce the law" (Michael Winerip, "Go On, Dump It Anywhere at All. Who's to Know?" *New York Times* [January 8, 1991], p. B1).

7. A New Cast of Mind

1. Emily Yoffe, "Silence of the Frogs," *New York Times Magazine* (December 13, 1992), p.64.

2. Ibid.

3. Ibid., p. 66. One likely culprit research has so far uncovered is "the rise in ultraviolet radiation caused by the thinning of the atmospheric ozone layer." The difficulty is that "a form of radiation in sunlight called ultraviolet B" is damaging the DNA of frog eggs beyond its natural capacity to repair itself, and the eggs are highly susceptible to such ultraviolet radiation because their eggs "are unprotected by shells or leathery covers." The frogs themselves may be subject to damage since "their skin is bare of feathers or hair" (Carol Kaesuk Yoon, "Thinning Ozone Layer Implicated in Decline Of Frogs and Toads," *New York Times* (March 1, 1994), p. B12). The *Times*, in an editorial, remarks that this finding "has to be of some interest to another bare-skinned species" ("What the Frogs Are Telling Us," *New York Times* [March 4, 1994], p. E14).

4. The example is drawn from Deborah Tannen, *That's Not What I Meant* (New York: Ballantine Books, 1986), pp. 5–6. I have added to the example what I take to be the spouse's thinking that would explain the actual reaction, which was the angry exclamation, "You're driving me crazy! Why don't you make up your mind what you want?" (ibid., p. 6). Tannen's book is filled with such examples. Cultural differences are an obvious source for them, for instance.

5. Philip J. Hilts, "Studies Say Soot Kills Up to 60,000 In U.S. Each Year," *New York Times* (July 19, 1993), p. A16.

6. Ibid.

7. Douglas W. Dockery et al., "An Association between Air Pollution and Mortality in Six U.S. Cities," *The New England Journal of Medicine*, Vol. 329, No. 24

(December 9, 1993), p. 1753. Footnotes 5 through 12 of this article contain references to eight other helpful similar studies that provide "further evidence for the conclusion that exposure to air pollution contributes to excess mortality" (ibid., p. 1759).

This article contains some of the evidence reported in the article by Hilts, "Studies Say Soot Kills Up to 60,000 In U.S. Each Year," and it is one of the studies to which Dr. Samet was responding.

One reason fine particles are thought more dangerous is that, it is hypothesized, when they are inhaled deeply into the lungs, they clog the lungs up, as it were, and so diminish "the ability of the lungs to push oxygen into the blood, causing coughing, shortness of breath, sneezing and asthma attacks" ("Study Ties Fouled Air to High Urban Death Rates," *New York Times* [December 9, 1993], p. B15).

8. Dockery, "An Association," pp. 1755 and 1756. The situation in six cities was investigated—Watertown, Massachusetts, Harriman, Tennessee, "specific census tracts of St. Louis," Steubenville, Ohio, Portage, Wisconsin, "including Wyocena and Pardeeville," and Topeka, Kansas (ibid., p. 1754).

9. Ibid., p. 1756.

10. E. Gallie, "Essentially Contested Concepts," *Proceedings of the Aristotelian Society*, Vol. LVI, New Series (1955–56), pp. 167–198.

11. Ben Bolch and Harold Lyons, *Apocalypse Not: Science, Economics and Environmentalism*, (Washington, D.C.: Cato Institute, 1993), p. 3.

12. Dixy Lee Ray, with Lou Guzzo, *Environmental Overkill: Whatever Happened to Common Sense?* (Washington, D.C.: Regnery Gateway, 1993). The title of the book expresses its point of view. But Ray is clear that (1) "problems should be proved to be real before we lavish money on them" (pp. ix–x), and (2) "it is neither reasonable nor prudent for major political decisions to be based on presumptions about issues in science, which, in the current state of knowledge, are still only hypotheses" (p. 6). Ray attributes (2) to the scientists at the Earth Summit, but the context makes it clear she believes it too. It is (1) and (2) that provide the support for the sort of position Dr. Samet holds regarding soot and mortality. They seem, prima facie, so compelling, I have been suggesting, that it may be difficult to see them as competing with anything at all. They must seem to Ray to be definitive of common sense and certainly of the scientific mind.

13. Ray thus refers to those who opt for sustainable patterns of production and consumption as preferring "the socialist approach to problems." (ibid. 5).

14. All the quotations in this paragraph are from Keith Schneider, "Second Chance on Environment: Lawmakers Are Seeking Uniform Approach on Risk vs. Cost," *New York Times* (March 26, 1993), p. A11.

15. The example that follows is borrowed from the many Dr. Jasper Shealy III has of designs that provoke errors. I have also made use of his concept of the ideal design, but, as I have noted, turning the concept of an error-provocative design on decision-procedures is my own idea, and he bears no responsibility for having provoked any errors in a philosopher—the sort, some might say, so designed as to be perfectly capable of making errors on their own, without any provocation at all. For further reading on this concept of an error-provocative design, though not

put using that phrase, see Henry Petroski, *To Engineer Is Human: The Role of Failure in Successful Design* (New York: Vintage Books, 1992), esp. pp. 85ff.

16. Once one begins to hunt for error-provocative designs, one finds them everywhere. The main library in Rochester, New York, had on its men's room door on the second floor the following handlettered sign: "Turn Knob to the LEFT," with "LEFT" underlined three times. One wonders how many people complained of the bathroom door's not opening before the sign was made.

17. Given the uncertainties in the political environment, members of Congress are not likely to risk acting in ways that, if successful, would maximize their chances of re-election, for such actions, if unsuccessful, can make them look like bad politicians to their constituents—unable to bring home the bacon. They rather act in whatever way will maximize those chances—usually conservatively, as I have noted (see David R. Mayhew, *Congress: The Electoral Connection* [New Haven: Yale University Press, 1974], p. 47).

18. See, in this regard, Claud S. Rupert's remarks about a "wait-and-see" response having its own costs, quoted in *The Analysis of Actual Versus Perceived Risks* (New York: Plenum, 1983), p. 319.

19. It need not, of course, for we could require that someone who wishes to invest in a chemical plant, for instance, provide all relevant information that no substantial harm will result to the environment. We tend not to raise such a high hurdle to investment at least in part because we appeal to other values, like ensuring more employment opportunities, to encourage such investments.

20. "Atomic Waste Reported Leaking in Ocean Sanctuary Off California," *New York Times* (June 7, 1990), p. A12.

21. "Cheap Chemicals and Dumb Luck," *Audubon*, Vol. 78, No. 1 (January 1976), p. 110.

22. Ibid., p. 117.

23. It is this cast of mind that lies behind concerns that the listing of species as endangered comes too late to help them. Species are listed as "endangered if they are at imminent risk of extinction," and the concern is that once the population of a species plummets to that level "and groups become fragmented and isolated from one another, the probability of local extinctions from factors like inbreeding or random disasters like fire and hurricanes can rise sharply" and the chances of recovery of the species are significantly reduced. The proper response is to intervene earlier in the process when a species is "clearly at risk" and work to save enough of the ecosystem to provide a better margin of safety (John R. Luoma, "Listing as an Endangered Species Is Said to Come Too Late for Many," *New York Times* [March 16, 1993], p. A12).

It is this cast of mind that Secretary Babbitt has displayed in working toward arrangements with developers that allow for saving enough habitat for a species to survive in exchange for development of the rest. The deal made with developers regarding the California gnatcatcher is a model of this sort of thinking: "developers, conservationists and Federal and state officials are working together to preserve an entire ecosystem, rather than haggling over emergency measures prescribed by the Endangered Species act to save single species" (Robert Reinhold, "Tiny

Songbird Poses Big Test For U.S. Environment Policy," *New York Times* [March 16, 1993], p. A1). The aim is to "set up a dozen preserves meant to save not only the gnatcatcher but also its full ecosystem, which is host to 30 or 40 other sensitive species" (Robert Reinhold, "California Environmentalists Cut a Deal, Hope for the Best," *New York Times* [March 28, 1993], p. E4). But the gnatcatcher plan is only one of the 75 or so in progress. They can be complicated. The plan "for a system of preserves in Austin and Travis County, Tex., for two songbirds and five cave invertebrates" got "hung up over the question of who will pay for acquiring and managing the land" (Carl Hulse, "Building Near Endangered Species," *New York Times* [December 28, 1993], p. B8). But what is being proposed is clearly preferable to either allowing laissez-faire development and then trying to handle the environmental problems that creates after the fact or prohibiting any development in areas that are sensitive. Holding the line against development requires a political commitment over many administrations and sessions of Congress that is not likely to be sustained. But an agreement to allow some development in exchange for enough protection to ensure the survival of a species is an agreement that can last.

24. Thus, from *within* what I am arguing is the proper cast of mind, it is quite appropriate to say, with Senator Moynihan, "We're entirely capable of moving on to a more productive period of environmental protection that is based on clear factual assessments of what produces risks to human health and the environment" (Schneider, "Second Chance on Environment: Lawmakers Are Seeking Uniform Approach on Risk vs. Cost," p. A11). No one denies the need for more research or for the desirability of basing decisions regarding the environment on as firm a factual footing as we can. What is objectionable is to suppose that we cannot intelligently and morally act when the footing is less than it could be, as in the "robust correlation" we have for soot and mortality.

25. The idea is to stem the tide against the increasing loss of water and water quality that is affecting the ecosystem of the Everglades (John H. Cushman, Jr., "Everglades Cleanup Agreement Fails," *New York Times* [December 19, 1993], p. A.28). "A cornerstone of the plan is the purchase of 40,000 acres of farmland to be converted into six artificial marshes, which would cleanse runoff water of the chief pollutant, phosphorus." The phosphorus "stimulates the growth of plants like cattails that choke out the native species eaten by fish and wildlife. The agreement also calls for increasing the amount of fresh water flowing into the Everglades by 25 percent, a measure intended to restore plant and animals species that are vanishing because of declining water tables." The plan would cost Florida sugar cane growers $322 million of clean-up costs, although "if they reduce the phosphorus contamination at a faster pace, the clean-up costs could be cut by as much as $90 million" (Larry Rohter, "U.S. and Florida Growers Reach Pact on Everglades," *New York Times* [July 4, 1993], p. A12).

Perhaps more significantly, the plan would also not only take 40,000 acres of farmland out of production but does not and, arguably, cannot include any guarantee that more will not be needed (John H. Cushman, Jr., "U.S. and Florida Lean on Sugar Producers to Restore Polluted Everglades," *New York Times* [January 16, 1994], p. A20). The difficulty is that the steps agreed on may not be enough and that, in any event, it is claimed, "on the farmland, drained by thousands of miles of ditches and canals, the mucky topsoil was deteriorated so badly that farming would eventually become virtually impossible" (Cushman, "Everglades Cleanup Agreement Fails,"). The best solution to that eventuality might be to flood "hundreds of thousands of acres of farmland" (ibid.). Given such uncertainties, the growers, or some of them, have had hesitations about the agreement they made. That the agreement, reached in July of 1993, unraveled in late 1993 is a sign of how difficult and delicate such negotiations are.

26. One measure of the progress is that "from 1955 to 1992 the peak level of ozone—one of the best indicators of air pollution—declined from 680 parts per billion to 300 parts per billion" (James M. Lents and William J. Kelly, "Clearing the Air in Los Angeles," *Scientific American* [October 1993], p. 32). But not all the news is good. "Southern California's air quality is [still] the worst in the U.S. Air pollution . . . reaches unhealthful levels on half the days each year, and it violates four of the six federal standards for healthful air—those of ozone, fine particulates, carbon monoxide and nitrogen dioxide" (ibid.).

Unfortunately, with that progress and in the face of those difficulties, the move to continue to wring further pollution out of the air is having a more difficult time. In harder economic times, it is harder to make the case for cleaner air when that runs against economic development. "The district backed off from trying to force the South Coast Plaza, a big shopping mall in Orange County, from charging a parking fee when the Costa Mesa city government balked," and instead of mandating significant decreases in pollution, a Regional Clean Air Incentives Market has been created, allowing for a "new market in pollution credits" that allow the "390 companies covered—those producing four or more tons a year of emissions—the flexibility to find the cheapest way to reduce emissions by trading pollution rights." There is a "declining cap on the total emissions," and such a market may be the best solution to the problems of trading off economic progress with progress on pollution, but the creation of the market is itself a sign of how much more difficult it now is to regulate cleaner air despite the need for Los Angeles to cut its ozone, for instance, "by half again to meet Federal and state clean-air laws" (Robert Reinhold, "Hard Times Dilute Enthusiasm for Clean-Air Laws," *New York Times* [November 26, 1993], p. A12).

27. Fresh water holds back the saline water that would otherwise back up into the delta. The Government proposes that, on the average, about 7 percent of the water that is now effectively drawn from the delta no longer be diverted. The harm to farms and cities that have come to rely on that water is obvious (Robert Reinhold, "U.S. Proposes to Divert Fresh Water to Save Imperiled Delta in California," *New York Times* [December 16, 1993], p. A18).

28. The aim in regard to the Florida Everglades is to "negotiate broad agreements with local interests based on mutual advantages." That the attempt collapsed makes it clear just how difficult it can be to accommodate environmental concerns with pre-existing economic interests (Cushman, "Everglades Cleanup Agreement Fails," p. A28).

29. In working out a compromise to protect enough of the habitat of the Cali-

fornia gnatcather to prevent its loss while allowing some development in the rest of its habitat, Secretary Babbitt was arguably working toward this end. Josel R. Reynolds, senior attorney for the Natural Resources Defense Council, said, "This is a major step in the right direction; Babbitt did provide something for everybody" (Robert Reinhold, "U.S. Acts to Save Home of Rare Bird," *New York Times* [March 26, 1993], p. A11).

30. This is a variant of Aristotle's remarks about how being moral is a difficult task. He says, for instance, that "it is not easy to determine in what manner, with what person, on what occasion, and for how long a time one ought to be angry" (*Nichomachean Ethics* [Indianapolis: Merrill, 1962], p. 51). This is a consequence of his view that in ethics "there are many ways of going wrong, but only one way which is right" (ibid., p. 43)—a view that may also hold for judgments involving such complexities as those I think we ought to take into consideration regarding environment matters.

31. It is certainly arguable, for instance, that our original reaction to the realization of how harmful asbestos can be was a mistake and that we often, though not always, risked more harm by trying to remove it from the environment than by leaving it in place, for "the removal often sent tiny asbestos fibers into the air" (Keith Schneider, "New View Calls Environmental Policy Misguided," *New York Times* [March 21, 1993], p. A30). The argument, it is important to note, is not that it is always wrong to attempt to remove asbestos but that the removal itself must be guided by a finer appreciation of the potential risks *in* removing it.

32. I would thus argue that Congress made a mistake in 1990 in the Clean Air Act when it did not act to limit "mercury emissions from coal-burning electric plants" (ibid.).

33. "When you are operating at the limits of what science knows, the big mistake would be to underestimate the real danger and leave people unprotected," says David D. Doniger, "a senior lawyer with the Natural Resources Defense Council" (ibid.).

34. Hilts, "Studies Say Soot Kills Up to 60,000 In U.S. Each Year."

35. In saying that not acting is both immoral and unreasonable, I speak from within the cast of mind I think appropriate, but I also say it after arguing that the mode of reasoning and way of thinking about environmental issues it is meant to displace is incoherent. If the dispute were simply between competing paradigms, we should not gain any leverage on it by criticizing from within each, but if one of the ways of thinking about environmental matters has been shown to be incoherent and to cause great harm, whereas the other is coherent and causes less harm, then it is no longer inappropriate to mark their differences in the way I just have.

INDEX

abortion, 35–36
accountants, 31–32, 187n53, 187n54, 187n55
acid rain, 3, 53, 65, 83, 144, 198n36, 209n56, 246n121
acids, 127, 169n17, 211n8, 229n45, 246n119
acting rationally: competing conceptions of, 36–37, 146, 152–56, 167n9. *See also* caution; harms; incoherence; information; showing of greater harm
Adirondack Council, 209n56
adjudication, 65–66, 175n9, 201n19
Agoes, Mary, 130
Agriculture Department, 120
AIDS, 166n8
air: 138–39, 241n106, 243n112; as toxic waste dump, 194n27. *See also* pollution
air conditioners, 168n12, 197n31
Alaska, 118, 126
Allan, Dr. J. David, 23, 182n32, 182n33
Allenwood, Pennsylvania, 14, 173n1, 213n11
American Revolution, xi, 24. *See also* revolution
Antarctica, 111, 112, 121, 225n25, 225n27
aquifers, 220n10, 221n12
Arctic, 225n25, 225n27, 226n32
Argentina, 226n28
Aristotle, 169n16, 253n30
Arizona, 217n23
Arthur Kill, 206n43
asbestos, 4, 167n9, 168n10, 171n24, 178n17, 207n45, 253n31
aseptic packages, 219n32
ash, 11

Asia, 226n28, 226n32
association with harm, 165n3. *See also* correlation
Atomic Energy Commission, 127
Austin County, Texas, 250n23

Babbitt, Secretary, 250n23, 253n29
Baca, Jim, 224n20
baghouse equipment, 39, 161–62
Bailyn, Bernard, xi, 24
Bakhtin, M. M., 200n9
bats, 25, 184n39
batteries, 11, 12, 19, 86, 99, 129, 130–31, 133, 134, 135, 141, 147, 218n28, 243n113. *See also* mercury
Baucus, Senator Max, 240n105
Bauer, Chief Judge William J., 170n20
Bazzaz, Fakhri A., 121
BBC, 227n34
Belgium, 215n15
Benedict, Ruth, 200n10
benefits. *See* front-loading
Benlate, 21, 180n26
Berman, 205n39
Biddeford, Maine, 95
BIV. *See* bovine immuno-deficiency
black holes, 63, 79, 120, 128, 149, 229n45, 233n63
Blanchard, Governor, xi
blowing the whistle, 188n60
bottle laws, 82, 83, 113, 141, 146, 240n105
bovine immuno-deficiency, 166n8
British Columbia, 208n50
British health-care system, 193n15
British Petroleum, 224n20
Broecker, Dr. Wallace S., 194n25
Bronx, 190n1, 243n114

256   Index

Brooklyn, 191n8
brown fields, 206n43
Browning-Ferris Industries, 217n23
bureaucracy, 6, 168n13
Bush administration, 127, 132, 201n20
butter, 194n22

cadmium, 131, 133, 243n113
California: 29, 82, 84, 158, 179n22, 182n32, 182n33, 192n14, 211n8, 231n49, 252n26; gnatcatcher, 250n23, 253n29; roach, 182n33, 246n120
Callicoon, New York, 220n10
Campbell Soup Company, 237n83
Canada, 65, 83, 99, 110, 208n50, 209n57
cancer, 222n15, 223n18
carbon dioxide, 120, 122–23, 194n20, 194n25, 231n48, 233n62, 238n97, 246n121
carbon monoxide, 198n34
Carson, Don, 242n109
Cashore, John, 208n50
Catskills, 89, 177n13, 221n11
cattle, 166n8
caution, 50–51, 52, 117, 160–61, 232n61. *See also* acting rationally
Cavell, Stanley, 186n47, 186n50
Center for Disease Control, 130
Center for Environmental Health and Injury Control, 130
CFCs. *See* chlorofluorocarbons; CFC-11; CFC-12
CFC-11, 227n33, 228n42
CFC-12, 226n32, 227n33, 228n42
Challenger, 125, 187n55
checkers, 44, 65, 193n18
checking one's checkbook, 38–39, 40
Cheer, 132
chemical treatment facility, 68
Chesapeake Bay, 144–45, 244n116, 244n117, 244n118
Chicago, 11, 20, 30, 65, 93
Chile, 226n28
chlorinated organic compounds, 219n3, 222n14
chlorofluorocarbons, 109, 111–14, 115–17, 119, 122, 129, 132, 135, 141, 143, 146, 149, 155, 156, 157, 168n12, 198n35, 223n19, 224n24, 225n27, 226n31, 227n36, 228n37, 228n38, 229n43, 229n45, 229n46, 239n100. *See also* CFC-11; CFC-12
choice mechanisms, 198n1
circuit boards, 135, 229n45, 229n46
Clean Air Act, 155, 206n41, 209n56, 253n32

Clinton administration, 200n12, 201n20, 232n60
clock, 38, 148
Coca-Cola, 131
Colorado, 185n45
Columbia River, 128
common resource, 138–40, 220n10
computers, 240n105
conceptual shift, 150–51
Congress, 106, 203n26, 204n32, 250n17, 250n23, 253n32
Connecticut, 170n21, 171n24, 172n25, 173n27, 185n45, 220n5, 242n110
consequences. *See* effects.
conspiracy, xi, 127
conspiratorial mode of thought, xi, 21, 116, 180n25
Constitutional Convention, 104n35
conversions, 173n2. *See also* purification-rite
Cook, Richard, xi
Coopers & Lybrand, 187n53
Cornwall, 11, 83
correlation, 152–53, 251n24. *See also* association with harm
cost overruns, 39. *See also* projections of costs
Costa Mesa, 252n26
cost-benefit analysis, 205n38, 205n39, 245n117
courtesy, 105, 175n9, 185n43, 220n6
crabs, 245n118
Crews, David P., 180n27
criminal sanction, 247n123
Croton, 177n13
Cryptosporidium, 20, 177n14, 178n17
cultural anthropologists, 61
Curtis, Banky, 182n32
cutting bushes by a stream, 21–22, 24, 103

daffodils, 103–104
Darwin, 25, 27, 185n42
DDT, 211n8
deciding not to decide, 45–46
decision-procedure, 36, 43, 51, 52, 53, 57–58, 79, 82, 102–103, 114–24, 128, 151–52, 162, 201n14, 211n8, 253n35. *See also* acting rationally
decision process: God's, according to Leibniz, 8; ours, 57–61, 199n2, 205n40. *See also* acting rationally; decision-procedure
Declaration of Independence, xi
deforestation, 191n7
degradation of the environment, 182n33

Delaware County, 177n13
democracy, 137
Denison, Richard, 237n80
dentists, 62, 80–81
Department of Agriculture, 166n8
Department of Environmental Conservation, 248n123
depth grammar, 60
DES, 222n14
designs: not the result of intention, 19–25, 180n25, 182n34, 184n39, 188n60; the result of intention, 16–17. *See also* intentional action
detergents, 131–32
Detroit, 39, 50, 88, 99, 183n35
diapers, 83, 98, 218n26, 218n27
Dibson units, 225n25
Dieffenbacher-Krall, John, 95
Dinkins, Mayor David, 177n13
dioxin, 107, 190n1, 222n15
dishes, 17–18. *See also* lead
Donahue, Jr., James J., 218n28
drift-net fishing, 192n13
drink boxes. *See* juice boxes
Dryzek, John S., 198n1
Duffy, Leo P., 127
dumping at sea, 171n23, 208n51, 212n9. *See also* the Gulf of the Farallones; *KarenB*; *KhianSea*; *Mobro*
dumping fees, 39, 83, 85–86, 191n8, 213n10
dumps, radiation, 35, 74
duPont, 116, 158, 168n12, 180n26, 224n24, 228n38
Duracell, Inc., 218n28
Dutch, 214n13
Dworkin, Ronald, 175n9

East Europe, 195n29
East Liverpool, Ohio, 194n27, 201n20, 217n23
eating, 46
EBDC, 223n18
economic: compensation, 68; interests, ix; system, 75
economists, 200n12
Eden, 1, 69
effects, unintended, 16–25, 57, 127, 180n27, 181n29, 182n33, 188n60, 191n8, 198n1, 214n14, 135n65. *See also* designs
Ellen, Eric, 171n23
Emergency Planning and Right to Know Act, 84
emission allowances, 197n32
endangered species, 250n23

Endangered Species Act, 155
Energy Department, 196n30, 235n71, 236n75
England, 11, 12, 24. *See also* Great Britain
English, 9, 30, 32
environmental: guru, 15; purification rite, 14
Environmental Defense Fund, 99, 214n14, 237n80
Environmental Protection Agency (EPA), 77, 113, 162, 168n12, 170n20, 177n14, 179n19, 179n20, 201n20, 219n32, 223n18, 224n24, 226n28, 228n37, 236n75, 240n105, 242n110, 246n121
Environmental Quality Institute, 179n22
EPA. *See* Environmental Protection Agency
epistemological: card, 38–41, 70, 80, 101, 119, 149, 159–60, 193n15, 208n50, 211n8; state, 38, 52
equal protection, 168n14, 175n9
error-provocative designs, 7, 8, 156–58, 169n15, 188n60, 249n15, 250n16. *See also* designs; Savannah River Site
Escherichia coli, 176n12
essentially contestable concepts, 36, 154–55, 168n14, 219n3
Europe, 83, 226n28, 226n32
European Economic Community, 12, 85
Eurotunnel, 192n9
Everglades, 160, 242n109, 251n25, 252n28
evolution, 185n46, 174n6, 174n8, 175n9. *See also* bats; Darwin; Saussure
expectations, rightful, 5, 49–50
experts making decisions, 81–82
extinction of wildlife, 36
Exxon, 233n63
Exxon-Valdez, 109, 118–20, 123, 124, 230n47, 231n49

fatty tissues, 47
faucets, 179n22
Fazio, Representative Vic, 192n14
FDA (Federal Drug Administration), 192n14
Federalist Papers, 63, 204n32
feline immuno-deficiency, 166n8
Finkel, Dr. Adam, 155
First Amendment, 64
fish, 232n61, 235n66
FIV. *See* feline immuno-deficiency
Flathead Lake, 181n29
Florida, 242n109, 252n28
Flo-Sun, 242n109
Food and Drug Administration, 165n1, 228n37

258   Index

Formosa Plastics, 246n119
Fourteenth Amendment, 168n14
Fowler, Senator Wyche, 192n14
France, 12
Franson, Wieger, 225n27
Freeland, Michigan, 169n17
French, 27, 33, 245n117
Fresh Kills, 191n8, 211n7, 213n10, 218n26
frogs, 150, 161, 248n3
front-loading: benefits, 63, 71–76, 78, 82, 126, 157, 205n40, 206n41, 211n6; burdens, 8
fungicide, 223n18. See also Benlate

gambling, 35, 42–44, 54, 55, 193n19
garbage barge, 169n17
gasoline, 198n34, 238n94
gathering information, 3–4
General Motors, 131
German, 245n117; student, 29
Germany, 84, 209n53, 240n105
Gerry, Mr., 204n35
giardia, 177n14
Gilbert and Sullivan, 186n48
global warming, 231n48, 234n64. See also greenhouse effect
God, 1, 6, 8, 22
Gore, Al, 203n28
Grand Banks, 232n61
Great Bear Spring Company, 220n10
Great Britain, 26, 60. See also England
greenhouse effect, 102, 120–24, 132–33, 146, 148, 149, 232n60, 238n88. See also global warming
Greenland, 226n32
Greenpeace, 219n3
Greenwood, Mark, 240n105
Gulf of the Farallones National Marine Sanctuary, 107, 126, 223n19, 235n66
Gulf of Monterey, 231n49

Halm, Howard L., 211n8
Hanford Reservation, 127–28, 236n73, 236n75, 237n76
Harlem, 243n114
Harmon, Dr. Harry, 236n73
harms, 105–106, 107–108, 115, 119, 129, 154, 188n60, 194n27, 196n30, 233n63, 234n64, 238n89, 253n35; catastrophic, 112, 232n61; risks when magnitude is great, 37, 224n20, 227n34, 227n36, 228n40, 236n73, 250n23; as setbacks to interests, 176n11, 199n4. See also acting rationally; association with harm; incoherence; showing of greater harm
Hart, H. L. A., 186n49

Harwood, Mary, 232n61
Haverstraw, 172n25
hazard, 179n18
hazardous waste. See also waste
HCFCs, 114, 116
health: care system, 16, 32, 107, 182n34, 185n41; insurance, 188n56, 188n58, 188n59. See also British health-care system
heavy metals, 39, 107, 126, 130–32, 133, 139, 149, 156, 194n27, 211n8. See also cadmium; lead; mercury
Hekman, Jr., Peter M., 196n30
herbicides, 184n40
High Bridge, New Jersey, 247n122
high fliers, 35, 43, 51, 78–79
highest standards: of proof, 193n15; of scientific rigor, 166n7
Hispanics, 188n58
Honhenemser, Christoph, 170n18
hormonal systems, 222n14
Household Hazardous Waste Collection Day, 170n21
Hudson River, 172n25, 197n31, 247n123
Hume, David, 22, 23, 25, 60, 61, 63, 73, 77, 95, 176n10, 182n31, 182n34, 242n108. See also stranger
Hungary, 195n29
hydrocarbons, 195n29
hydrogen sulfide, 172n25

Idso, Sherwood B., 120
Illinois Power Company, 206n41
incinerators, 4, 11, 24, 35, 39, 50, 52–53, 67–68, 69, 77, 88–96, 99, 105, 139, 183n35, 190n1, 194n27, 201n20, 203n24, 206n43, 209n53, 213n11, 213n12, 215n16, 216n19, 217n23, 217n24, 217n25, 219n32, 220n5, 239n99
incoherent vision: of how to make a rational decision, 4, 10, 13, 37, 41, 48, 146, 154, 159, 161, 174n4, 190n3, 253n35; of waste management, 134
information: artifically limit amount of, 193n19; do not need all relevant, 41–47, 54–56, 146, 154, 233n63, 234n64; need all relevant, x, 2–4, 8, 10, 36–37, 38, 40–41, 42, 48, 79, 113, 148, 153–54, 227n36, 250n19. See also acting rationally
inoculation, self-, 58, 76–82
Institute for Local Self-Reliance, 220n5
insurance, 132–33, 238n87
intentional action, 175n7, 180n25, 188n60, 198n1. See also designs

interest-resolution, 63–69, 70–71
interests: 239n102, 245n117; balancing, 135, 205n39, 232n61; defined, 105; distinguished from wants, 105; distinguished from harms, 105–106, 110; as lines of force, 204n32; weighed against rights, 135–37, 201n14, 201n18. *See also* harms
internal point of view, 29–31
International Maritime Bureau, 171n23
Iran-Contra, 193n19
Irvine, California, 82
Italy, 192n13, 208n51

Jack-in-the-beanstalk, 43
Jackson, Dr. Jeremy B. C., 231n49
Japan, 1, 192n13, 229n45, 235n66
Johnson & Johnson, 214n14
joke about the Englishman, 18, 28, 154
judicial systems, 33–34, 201n20
juice boxes, 100, 219n32, 244n115
justice, 176n10, 242n108

Kaiser Engineers worker, 237n76
Kant, 187n52
KarenB, 12, 208n51
Kearney, Thomas A., 211n8
Kemp's Ridley turtle, 180n27
Kentucky, 217n23
Kessel, Richard M., 212n9
ketchup, 99–100
Khian Sea, 169n17
king of England, xi, 24
knowledge. *See* acting rationally; information
Kramer, Ken, 221n12
Kreiger, Andrew J., 220n10
krypton-85, 198n35

laissez-faire: decision making, 6, 11, 37, 55, 157; development, 250n23
Lake Michigan, 190n2
Lamar, Colorado, 185n45
land use, 21–23, 24, 25–26, 103, 182n32
landfilling, 90–91, 124, 128, 183n36, 191n8, 209n53, 211n6, 211n7, 213n10, 213n12, 214n15, 215n16, 216n19, 235n65
language, as natural social artifact, 9, 17, 18–19, 26–27, 28–30, 175n8, 185n42, 186n47
law, 11, 20, 27, 30, 39, 65, 81, 82, 83, 129, 138, 155, 157, 172n25, 179n22, 183n35, 186n51, 197n32, 201n14, 201n20, 205n39, 207n49, 209n53, 209n56, 209n57, 214n14, 220n5, 239n99, 240n104, 242n110, 247n123, 252n26; economic analysis of, 182n34, 189n61
lawns, 22, 107, 143, 184n40
"le weekend," 27, 33
lead: in batteries, 12, 19; in dishes, 107, 223n17; in dumpsites, 211n8; in fill, 171n24; in gasoline, 226n32; in water supply, 20–21, 22, 65, 129, 149, 170n18, 178n18, 179n19, 179n20, 179n22, 215n16. *See also* heavy metals
legal system, 175n9, 186n49
Leibniz, 8
Leicester, 17
Levy, Paul, 178n15
liability for pollution, 172n25
Liege, Belgium, 215n15
light of reason, 14
Lilco, 209n56
Lindzen, Richard S., 120
linguistic philosophy, 186n47
logging, 233n63
Long Island, 209n56, 239n99
Long Island City, 78
Long Island Sound, 145n118
Los Angeles, 69–70, 160, 252n26
Louisville, 25

McDonald's, 214n14, 242n111
Madison, James, 63, 204n32
Maine Energy Recovery Company, 95
Maine People's Alliance, 95
Manhattan, 77, 176n12, 243n114
margarine, 194n22
Massachusetts, 24, 172n25, 178n15, 195n28, 249n8
Massi, Michael J., 211n7
maximax rule, 43, 51,
maximin rule, 43, 51, 53–54
measuring to the limits of instrumentation, 110, 165n1
Medicare, 188n56
mercury, 39, 47, 99, 107, 126, 130, 133, 194n27, 218n28, 223n19, 232n55, 237n78
Mescalero Indians, 202n21
methane, 121
Miami, 183n36
Michigan, 82, 113, 124–25, 126, 130, 169n17, 237n78
Michigan Chemical Company, 158
Michigan Department of Agriculture, 1, 2, 109, 165n1
Michigan Department of Public Health, 1, 2, 76, 165n3
Michigan Farm Bureau, 48

Michigan Toxic Substance Control Commission, xi
micro-natural social artifacts, 25, 26–27
milk, 26, 165n3, 194n22; in disposable pouches, 30, 83, 98, 99, 216n17
Mill, John Stuart, 186n51, 187n52
Milwaukee, 20, 178n15, 178n17
miscommunication, 151, 219n3
Mitsubishi, 73
Mobro, 169n17, 208n51
Montana, 181n29
Moore, Allen, 87
moral: issues, 32–34, 49–50, 71, 188n60, 189n61, 193n19; relations, 240n104
mortality rates, 188n57
Moynihan, Senator Daniel Patrick, 155, 251n24
Muscott Reservoir, 177n13
Mysis relicta, 181n29

Namias, Jerome, 120
NASA, 187n55, 224n20
Nassau County, 212n9
National Solid Waste Management Association, 87
Natural Resources Defense Council, 253n29
natural social artifact: as analytical device, 220n8; as artifact, 15–16; assessing, 103–107, 189n61; determinative of how interests are conceived, 105; with ends in tension, 220n4, 245n117; as evolving, 174n6, 220n6; includes a decision-procedure, 9, 31, 34, 35; as natural, 15–19; as normative, 28–32; pervasive, 18, 207n48; privileged, 148–49; relation to intentions, 16–17, 19–25, 188n60, 235n65; relation to morality and values, 32–34, 186n49; as responsive to natural needs, 16; responsive to variety of causal factors, 16, 27, 88; scope, 19, 25–28; subject to only certain kinds of change, 9, 244n115. *See also* micro-natural social artifacts; effects; and for examples: courtesy; dishes; land use; language; lawns; legal system; political process; smoking; transportation system; waste management; water supply
Needleman, Dr. Herbert, 179n19
nesting boxes, 180n27
Netherlands, 12, 214n14
New Jersey, 206n43, 209n57, 220n5, 247n122, 247n123
New York City, 85–86, 89, 172n25, 176n12, 177n14, 178n17, 185n45, 191n8, 197n31, 206n43, 211n7, 213n12, 214n14, 221n11, 235n65, 240n105, 243n114
New York State, 77, 207n49, 209n56, 213n10, 220n10, 221n11, 237n81, 239n99, 246n120, 247n123, 250n16
New York State Consumer Protection Board, 212n9
newborns, 188n59
newspapers, 50, 86, 183n36, 214n14, 218n26, 235n66
Nigeria, 12, 83, 208n51
NIMBY, 67–69, 203n24
nitrogen oxides, 198n34
norms, 186n49, 186n50, 186n51, 187n52
North America, 126, 226n28
North Hempstead, 239n99
North River, 197n31
Northern Hemisphere, 225n27, 226n28, 226n32, 228n42
Nouwen, Paul, 214n14
Nowell, Major Charles, 247n122
nuclear plants, 12

objectivity, 2–3, 22, 207n47. *See also* stranger
obligations, 240n104
Office of Pollution Prevention, 240n105
OH, 121, 123
Ohio, 191n8, 194n27, 210n59, 213n10, 222n13, 249n8
Oil Pollution Act, 230n47
oil spill, 205n39, 230n47, 231n49, 233n63, 247n123
oil tankers, 73–74
oil wells, 221n12
Olver, Richard L., 224n20
omissions, 17, 18, 29, 48, 174n5, 194n21
omniscience, 2–3, 38
Onondaga Lake, 172n26
opinion, punished by, 186n51
Orange & Rockland, 246n120
oystering, 244n116, 245n118
ozone, 252n26; depletion, 111–12, 113, 117, 157, 194n25, 224n24, 225n25, 225n26, 225n27, 226n28, 227n33, 227n36, 228n38, 228n41, 228n42, 248n3

Palisades Nuclear Plant, 190n2
Panama, 231n49
paper, 214n14
paradigms for guidance, 101, 109–14, 186n50, 253n35
passing the buck, 83, 86

PBB. *See* polybrominated biphenyl
PCB, 1, 171n23
PCDDs, 47
PCDFs, 47
Pennsylvania, 14, 173n1, 191n8, 213n10, 212n11
phosphorus, 251n25
photograph, 16–17
physicians, 45–46
Pimentel, Dr. David, 203n30
Pine River, 158
points of view: external, 29; internal, 29–31
political: paralysis, 3; process, 57–58, 86. *See also* decision process
politicians, acting as politicians, 6, 7, 58, 59–63, 64, 67, 69, 73, 75, 76, 77–79, 81, 83, 95, 160, 193n15, 200n12, 200n13, 206n43, 212n9, 250n17
pollution: 75, 84, 106, 206n43, 233n63, 247n123; air, 11, 84, 106, 107, 139, 201n20, 204n32, 206n41, 209n56, 226n32, 241n32, 243n112, 252n26, 248n7; auto, 195n29; ground-water, 24, 90, 107, 198n35, 221n13, 222n13, 236n73, 238n94; oil, 73, 221n12, 230n47; organic and chemical, 182n33; particle, 3, 152–53; water, 144, 146n119, 177n13, 243n112, 244n109
pollution control devices, 39, 183n35
pollution credits, 206n41, 209n56, 252n26
Poly Peck International, 187n53
polybrominated biphenyl (PBB), ix, 1–2, 3, 40, 48, 55, 76, 78, 80, 109–10, 112–13, 117, 143, 158, 165nn2 & 3, 166n4, 194n26, 211n8
polychlorinated: dibenzodioxins, 47; dibenzofurans, 47
polyethylene terephthalate (PET), 99, 219n32
pooper-scooper laws, 27
Portney, Kent E., 203n24, 206n43
practice: of doing the dishes, 17–18; of land use, 21–23; reasonable, 15; systemic features, 9
precedent: 49, 51, 212n9; in the law, 5, 156, 172n25
precision, 10, 169n16
predictability, 47, 194n25, 223n19
predicting, 228n42, 234n64. *See also* projections of costs
presuming: full knowledge, 2–3, 166n8; rationality, 190n3
presumptions, 37, 53, 124–29, 191n7, 222n15, 236n73, 237n76
Prince William Sound, 118, 119, 123, 124, 128, 132, 158, 162, 230n47, 231n49, 233n63
privacy, 27, 33, 185n44
Proctor & Gamble, 218n27
professionals, 58–62, 79, 138, 187n55, 199n3, 204n34, 240n104
projections of costs, 192n9, 197n31, 215n16, 217n23. *See also* predicting
prophets, 173n3
provoking errors. *See* error-provocative designs
prudence, 71
public: good, 7, 62, 77, 81, 160, 168n14, 247n123; goods, 144
public, not rational, 174n4
public policy: decisions, 2, 3, 9, 15, 36, 37, 38, 39, 40, 49, 51, 53, 62, 64, 75, 78, 80, 82, 91, 122, 153, 159, 161, 162, 198n1, 232n58; ethical issues in, 188n60; form of discussions about, 118, 119, 129, 150, 152; major decisions in, 3, 152, 159; matters, 41, 50, 95, 145, 148, 154, 207n47, 219n3; setting, 6, 121, 191n8, 233n63. *See also* acting rationally; decision-procedures; decision process, effects; front-loading; incoherent vision; political paralysis; precision; re-election, showing of greater harm
public property, 138–40
Puerto Rico, 205n39, 230n47, 247n123
Punta Arena, Chile, 226n28
purification rite, 14. *See also* conversions

quality trash, 171n22
Quayle, 77

radiation, as health hazard, 168n11, 248n3
radiation dumps, 35
radioactive fuel, 235n71
rainbow, 151, 152
rational action. *See* acting rationally
Rawls, John, 168n14
Ray, Dixie Lee, 122, 225n27, 249n12
Raymark Industries, 171n24
reactor, 196n30
recycling, 92, 94–95, 99–100, 131, 171n22, 183n36, 207n49, 209n53, 214n13, 214n14, 215n15, 217n25, 219n32, 220n5, 238n94, 240n105, 242n111, 243n114, 247n123
Red Dye No. 3, 192n14
re-election, 6, 62–63, 72, 200n13, 203n26, 204n33, 250n17

refrigerator, 112, 226n31, 228n40, 229n45, 229n46
Regional Clean Air Incentives Act, 252n26
Reilly, William K., 226n28
relations, 240n104
RemTech, 190n1
Resource Conservation and Recovery Act, 155
Resources for the Future, 155
responsibility, 69, 76, 84, 193n19, 195n29, 233n63, 240n105, 249n15
revolution, 105
Revolutionary War, 24
Reynolds, Josel R., 253n29
Rhode Island, 220n5, 246n120
rightful expectations, 5
rights, 62, 64–67, 68–69, 135–43, 144, 201n14, 201n18, 201n20, 232n61, 239n102, 239n103
risk: assessments, 35, 155, 190n2, 190n3, 191n4, 191n8, 203n24, 222n15, 224n23, 233n63; averse, 197n32; substitution, 68, 203n24. *See also* harms; taking a risk
rock climbing, 43
role-morality, 60–61
Romans, 60
Rousseau, xi
rules for decision making. *See* decision-procedures; decision processes; maximax; maximin
Rupert, Claud S., 227n36, 250n18
Ruston, John F., 99, 215n14

sablefish, 235n66
Sacramento Delta, 160
Sacramento River, 182n32
Sacramento sucker, 182n33
Sahara, 75
salmon, 181n29
Salvation Army, 247n122
Samet, Jonathan, 3, 152–54, 160, 166n6, 248n7
San Andreas fault, 126
San Antonio Bay, 246n119
San Francisco Bay, 160
San Juan, 230n47, 247n123
Saussure, Ferdinand de, 185n42
Savannah River Site, 196n30, 236n75
savings and loan institution, 31, 199n8
Schwartz, Dr. Joel, 161
scientific method, 2–3, 79–80, 148, 249n12
Scripps Institution of Oceanography, 120

scrubbers, 47, 99, 197n32, 206n41
Selman, Neil S., 220n5
Senate Committee of Environment and Public Works, 155
Shapiro, Fred, 237n81
Shealy III, Jasper, 168n15, 249n15
Sheehan, John, 209n56
Sherman, Mr., 205n35
Shoreham, 216n19
showing of greater harm, ix–x, 2, 8, 101, 121–22, 158
shrimping, 246n119
Sierra Blanca, 185n45
Sinclair, Dr. Mary P., 190n2
slavery, 204n32
sludge, 185n45, 212n9, 238n92, 247n123
Smith, Adam, xi
smog, 203n27
smoking, 17
social artifact. *See* natural social artifact
socialism, 249n13
soot, 3, 152–54, 158, 159, 161–62, 248n7. *See also* pollution; particles
South Coast Plaza, 252n26
Southern Hemisphere, 225n27, 226n28, 232n55
Southern Ohio Coal Company, 210n59
Soviet Union, 75, 127, 226n28, 226n32, 227n34
soybean ink, 133, 238n92
space shuttle, 224n20
Spanish, 27
Spies, Dr. Robert B., 233n63
Stamp Act, 24
statements, internal and external, 186n49
Staten Island, 218n26, 243n114
status quo, privileging, ix–x, 3, 40–41, 101–102, 146, 162
stop sign, 29
stove burner, 156
stranger, objective reasonable, 22, 23, 24, 29, 103, 182n31, 182n34, 198n1
Stringfellow, 211n6, 211n8
Subcommittee on Public Health and Environment, 224n24
sulfur dioxide, 198n36, 246n121
Sununu, John, 231n48
Superfund Law, 155, 173n27, 242n110
Supreme Court, 11
swans, 177n13
Symington, Governor Fyfe, 217n23
Syracuse, 172n26
systemic features, 9, 201n15

taking a risk, 107–109

tankers, 73–74, 205n38
terms of criticism, 30, 186n50
Terre, 215n15
Tetra Pak, Inc., 219n32
Texas, 185n45, 246n119, 250n23
Texas Sierra Club, 221n12
Thomaston, Connecticut, 173n27
Three Mile Island, 174n4
Time, 214n14
tires, 147, 240n105
topsoil, 70
Townsend Duties, 24
Trans Alaska Pipeline, 224n20
transmogrification, 65–66, 201n20
transportation system, 88, 136
trash, 4, 11–14, 19, 23, 28, 48, 54, 63, 77, 83–85, 89, 90–93, 95, 133, 134, 147, 169n17, 170n21, 171n22, 183n35, 183n36, 191n8, 195n28, 211n7, 212n9, 213n10, 213n12, 214n13, 214n14, 215n15, 217n24, 217n25, 239n99, 242n111, 243n114, 244n115, 247n122
Travis County, Texas, 250n23
tritium, 198n35
trust, 50, 196n30
turtle, 180n27

Ueber, Edward, 223n19
unintended effects. *See* effects
United Nations, 192n13
United States, 26, 27, 29, 65, 83, 84, 85, 96, 99, 112, 113, 133, 166n8, 204n31, 208n50, 214n14, 218n26, 222n13, 231n48, 238n94, 243n113
University: of New Mexico, 152; of North Carolina, 179n22
unscientific, 148. *See also* information
U.S. Gypsum, 172n25
Ushusiah, Argentina, 226n28
Utah, 173n1
utilities, 197n32, 209n56, 229n46, 246n120, 246n121

Valdez, Port of, 231n49
value judgments, 32–34, 78–79, 207n47
vetiver, 70–71, 204n31
Vial, James, 150

Victoria, 208n50
vitamin D, 194n22
Vogelsburg, F. Anthony, 116, 228n38

Warren-Washington Counties Industrial Development Agency, 239n99
Wartburgs, 195n29
Washington: D.C., 220n5; State, 127, 128, 168n11, 236n75
waste: 1, 240n105; amount produced, 87, 96, 239n98; hazardous, 11–13, 84–85, 140–41, 170n18, 170n21, 209n57, 217n23, 238n92; industrial, 171n22; medical, 190n1; necessary, 47; nuclear, 202n21, 236n73; oil, 209n57; radioactive, 11–12, 127–28, 190n2, 237n76; requires an expert, 80–81; as resource, 27, 88, 185n45, 239n99; solid, 11, 12, 170n21, 244n115; toxic, 194n27, 242n110. *See also* water
waste management: conditions for a practice, 23; current practice, 4, 5–6, 23–24, 34, 89, 97; as natural social artifact, 16, 94, 102, 105, 134–35, 151–52, 157, 174n6, 216n20, 220n4
water: bottled, 220n10; ground, 198n35; as resource, 138–39; supply, 19–21, 106, 138–39, 170n18, 176n12, 177n13, 178n15, 178n16, 178n17, 179n18, 179n20; waste, 246n119. *See also* lead; pollution (ground-water & water)
watershed, 139, 177n14
Watkins, Secretary, 196n30
Westinghouse, 236n73, 236n75
Whiffle Balls, 238n96
White House, 77, 194n20, 231n48
Whirlpool, 228n40
Wibbels, Thane R., 189n27
Williams, Michael, 208n50
Willis, James D., 239n99
Wittgenstein, 186n47
Wood, Gordon S., 220n6
wood ducks, 180n27
World Wildlife Fund, 36
worst-case scenarios, 119, 125–26
wrongs, 32, 101, 136–37, 239n101

UNIVERSITY PRESS OF NEW ENGLAND publishes books under its own imprint and is the publisher for Brandeis University Press, Brown University Press, University of Connecticut, Dartmouth College, Middlebury College Press, University of New Hampshire, University of Rhode Island, Tufts University, University of Vermont, and Wesleyan University Press.

Library of Congress Cataloging-in-Publication Data

Robison, Wade L.
Decisions in doubt : the environment and public policy / Wade L. Robinson.
   p. cm. — (Nelson A. Rockefeller series in social science and public policy)
   Includes bibliographical references and index.
   ISBN 0-87451-695-1
   1. Environmental policy—United States.   I. Title. II. Series.
HC110.E5R625   1994
333.7—dc20   94-20549
∞